Internet of Things

Technology, Communications and Computing

More information about this series at http://www.springer.com/series/11636

Raffaele Gravina · Carlos E. Palau
Marco Manso · Antonio Liotta
Giancarlo Fortino
Editors

Integration, Interconnection, and Interoperability of IoT Systems

Springer

Editors
Raffaele Gravina
Department of Informatics, Modeling, Electronics, and Systems Engineering
Università della Calabria
Rende (CS), Calabria
Italy

Carlos E. Palau
School of Telecommunications Engineering
Universitat Politècnica de València
Valencia
Spain

Marco Manso
Rinicom
Lancaster
UK

Antonio Liotta
Department of Electrical Engineering
Technische Universiteit Eindhoven
Eindhoven, Noord-Brabant
The Netherlands

Giancarlo Fortino
Department of Informatics, Modeling, Electronics, and Systems Engineering
Università della Calabria
Rende (CS), Calabria
Italy

ISSN 2199-1073 ISSN 2199-1081 (electronic)
Internet of Things
ISBN 978-3-319-61299-7 ISBN 978-3-319-61300-0 (eBook)
DOI 10.1007/978-3-319-61300-0

Library of Congress Control Number: 2017945218

Printed on acid-free paper

This Springer imprint is published by Springer Nature
The registered company is Springer International Publishing AG
The registered company address is: Gewerbestrasse 11, 6330 Cham, Switzerland

Preface

IoT is an emerging concept that involves a larger and larger number of heterogeneous smart everyday-life objects. They enable a large scope of new applications that require all these objects to communicate, to interact, and to share data and processes. All these objects have popped up from there to there, using their own communication means, OS or language, resulting in a jeopardized ecosystem in which different solutions are often isolated and non-interoperable. A key open issue to realize the full capacity of the IoT is thus interoperability.

After some market consolidation happening mostly between 2011 and 2013, IoT products are now hitting the market across all segments: consumer (wearables, home automation), commercial (HVAC, parking), and industrial (industrial process control, supervision).

Often driven by the fear to "fall behind," small and large companies push their engineering teams to productize solutions quickly. If those companies choose to implement standards-based products, the compliance testing, interoperation testing, and labeling of their product may take over a year, which is often unacceptable giving today's rush-to-market. Companies therefore often go for in-house proprietary solutions, which can be developed and tested much faster. The result is that the market is highly fragmented: a large number of non-interoperable solutions are being installed, eventually leading to increased cost, inefficiencies, customer frustration, and a rate of adoption of the IoT much slower than the numbers touted by analysts.

The market is now at a state where we need to think about integration, interconnection, and interoperability. What does it take to make different IoT solutions seamlessly integrate with one another? Are there architectures and tools one could develop to speed up interoperability testing? If interoperability is not feasible or desired, can we at least build in mechanisms for a different product to coexist?

The main objective of this edited book is, hence, to try answering such questions by investigating the lack of interoperability in the IoT realm, including innovative research as well as technical solutions to interoperability, integration, and interconnection of heterogeneous IoT systems, at any level. The book also explores issues caused by lack of interoperability such as impossibility to plug non-interoperable IoT devices into heterogeneous IoT platforms, impossibility to develop IoT applications exploiting multiple platforms in homogeneous and/or cross domains, slowness of IoT technology introduction at large scale: discouragement in adopting IoT technology, increase of costs, scarce reusability of technical solutions, and difficulty in meeting user satisfaction.

The book is structured into 10 chapters focused on the above mentioned topics and providing novel and cutting-edge contributions for the next-generation IoT systems. A short introduction to the chapters is provided below.

Chapter "3GPP Evolution on LTE Connectivity for IoT", by Almudena Díaz Zayas and Cesar A. García Pérez and Álvaro M. Recio Pérez and Pedro Merino focuses on the provision of wide area and efficient connectivity to the Internet of Things (IoT), a key factor in such an explosion, through the usage of LTE. LTE MTC (machine type communication), LTE M2M, or just LTE-M are the coined terms that refer to this issue. The chapter also provides a detailed analysis of the standardization efforts carried out by the 3GPP to convert LTE into an IoT capable technology.

Chapter "Towards Efficient Deployment in Internet of Robotic Things" by Cristanel Razamandimby, Valeria Loscri, and Anna Maria Vegni proposes IoT-based, and a neural network control scheme to efficiently maintain the global connectivity among multiple mobile robots to a desired quality-of-service (QoS) level.

Chapter "Transmission Power Control in WSNs: From Deterministic to Cognitive Methods" by Michele Chincoli and Antonio provides an overview of the Transmission Power Control (TPC) protocols present in literature, categorized as deterministic (proactive and reactive) and cognitive (Swarm Intelligence, Fuzzy Logic, and Reinforcement Learning). The review identifies key shortcomings in deterministic TPC, pinpointing the benefit of the emerging methods based on computational intelligence.

Chapter "Devices Can Be Secure and Easy to Install on the Internet of Things" by Roger D. Chamberlain, Mike Chambers, Darren Greenwalt, Brett Steinbrueck, and Todd Steinbrueck describes the industrial deployment experience of the EZConnect™ IoT security infrastructure implemented by BECS Technology, Inc., a firm that provides water chemistry monitoring and control equipment to the aquatics market.

Chapter "A Service-Based Approach for the Uniform Access of Wireless Sensor Networks and Custom Application Tasks Running on Sensor Nodes" by Theodoros Fronimos, Manos Koutsoubelias, Spyros Lalis, and Thomas Bartzanas proposes an approach for supporting both application-specific sensing and processing tasks information flows simultaneously and in a uniform way, via an open Web-based service interface that can be flexibly extended through appropriate XML descriptions.

Chapter "Towards Semantic Interoperability Between Internet of Things Platforms" by Maria Ganzha, Marcin Paprzycki, Wiesław Pawłowski, Paweł Szmeja, and Katarzyna Wasielewska is focused on semantic interoperability, which can be seen as the meta-level for all interoperability considerations. The aim of this chapter is to consider how multiple IoT platforms can "understand" each other and have meaningful "conversation."

Chapter "Linked Data for Internet of Everything" by Danh Le-Phuoc and Manfred Hauswirth proposes the idea of "Linking Everything" by extending Linked Data Principles to interlink "everything" into a hypergraph to tackle the problem of interoperability in the next generation of Internet, namely "Internet of Everything."

Chapter "Sensing Enabled Capabilities for Access Control Management" by Mikel Uriarte, Oscar López, Jordi Blasi, Oscar Lázaro, Alicia González, Iván Prada, Eneko Olivares, Carlos E. Palau, Miguel A. Portugués, and Alejandro García presents a new framework approach that provides flexible, open, fluid, and collaborative middleware for building access control management systems, based on the sensing enriched access control (SEAC) concept. The chapter also provides a description of a real use case raised to validate the framework, as well as the laboratory results supporting its scalability.

Chapter "The Application of Telematics and Smart Devices in Emergencies" by Marco Manso, Barbara Guerra, Cosmin Carjan, Evangelos Sdongos, Anastasia Bolovinou, Angelos Amditis, and David Donaldson shares two interesting use cases depicting emergencies where the use of telematics and smart devices enable improved emergency situational awareness for citizens and emergency services.

Chapter "Towards Multi-layer Interoperability of Heterogeneous IoT Platforms: The INTER-IoT Approach" by Giancarlo Fortino, Claudio Savaglio, Carlos E. Palau, Jara Suarez de Puga, Maria Ganzha, Marcin Paprzycki, Miguel Montesinos, Antonio Liotta, and Miguel Llop discusses how IoT interoperability is addressed within the INTER-IoT Project. The INTER-IoT voluntary approach supports the creation of new IoT interoperable ecosystems with hardware/software tools, frameworks for open IoT application and system programming and deployment, and engineering methodology for IoT systems integration.

We would like to thank all the book contributors, the anonymous reviewers, and Prasanna Kumar Narayanasamy from Springer for the support during the publication process.

Rende, Italy	Raffaele Gravina
Valencia, Spain	Carlos E. Palau
Lancaster, UK	Marco Manso
Eindhoven, The Netherlands	Antonio Liotta
Rende, Italy	Giancarlo Fortino

The original version of the book was revised: Incorrect author name has been corrected. The erratum to the book is available at https://doi.org/10.1007/978-3-319-61300-0_11

Contents

3GPP Evolution on LTE Connectivity for IoT

Almudena Díaz Zayas, Cesar A. García Pérez, Álvaro M. Recio Pérez and Pedro Merino

Abstract Nowadays, the maturity of IoT applications, networking technologies and manufacturers of "things" have caused an explosion in the number of connected devices. There are several reports which calculate that the number of connected things will reach 50 billion in the near future. This chapter focuses on the provision of wide area and efficient connectivity to the Internet of Things (IoT), a key factor in such an explosion, through the usage of LTE. LTE MTC (Machine Type Communication), LTE M2M or just LTE-M are the coined terms that refer to this issue. This chapter provides a detailed analysis of the standardization efforts carried out by the 3GPP to convert LTE into an IoT capable technology.

1 Introduction

According to [1], by 2020 the number of connected devices, other than cell phones, PCs and tablets, will be 15 billion, of which 13.0 billion will use short range technologies, such as Bluetooth Smart, Wi-Fi or Zigbee, and 2.0 billion will use a cellular connection. However this report estimates that 5.5 of the 13.0 billion would be replaced by a cellular connection if that cellular technology could meet the requirements of the IoT [2]. Moreover, in that case, the incursion in the IoT world of a cellular IoT capable technology would add 5.7 billion of connections, reaching 20 billion of things by 2020. With these numbers, mobile operators and manufacturers have found in IoT a promising market, bringing with it a myriad of promising business opportunities. In fact, IoT devices are able to use PAN, LAN, WAN or cellular

A. Díaz Zayas (✉) · C.A. García Pérez · Á.M. Recio Pérez · P. Merino
Andalucia Tech, Edif. Ada Byron, University of Málaga, Málaga, Spain
e-mail: almudiaz@lcc.uma.es

C.A. García Pérez
e-mail: garciacesaraugusto@lcc.uma.es

Á.M. Recio Pérez
e-mail: amrecio@lcc.uma.es

P. Merino
e-mail: pedro@lcc.uma.es

R. Gravina et al. (eds.), *Integration, Interconnection, and Interoperability of IoT Systems*, Internet of Things, DOI 10.1007/978-3-319-61300-0_1

networks. Furthermore, the most restrictive IoT device requirements (battery powered, low cost and high autonomy), have fostered the use of wireless technologies such as RFID, NFC, Bluetooth or Zigbee.

However, as reported by [1], the adoption of cellular connections in the world of IoT could be a turning point, providing a significant increase in the number of connected things, which will turn cellular technology into the dominant networking option in the IoT. Nevertheless, in order to reach these numbers, mobile broadband networks have to evolve to become compatible with IoT.

Any existing cellular network could be chosen, such as GSM, UMTS or LTE. Until now, GPRS or CDMA2000 have been the technologies chosen to provide wide-area connectivity because they provide higher coverage and lower cost solutions. However, LTE has the greater potential to shake up the Internet of Things. The flat and flexible architecture of LTE fits perfectly in IoT ecosystems. Moreover, the efficient use of the spectrum of LTE directly translates into lower operating costs, which is key to provide global and cost-effective data connectivity. The original design of LTE did not address all the needs of IoT. However, according to the roadmap provided in [3], 4G solutions based on LTE Releases 12 and 13 will improve LTE capabilities so as to support IoT connectivity requirements.

Although LTE can be a great catalyser, understanding IoTs specific requirements is of utmost importance to drive this change. Figure 1 provides a picture of all the factors related to cellular technology dimensioning, and of the differences between

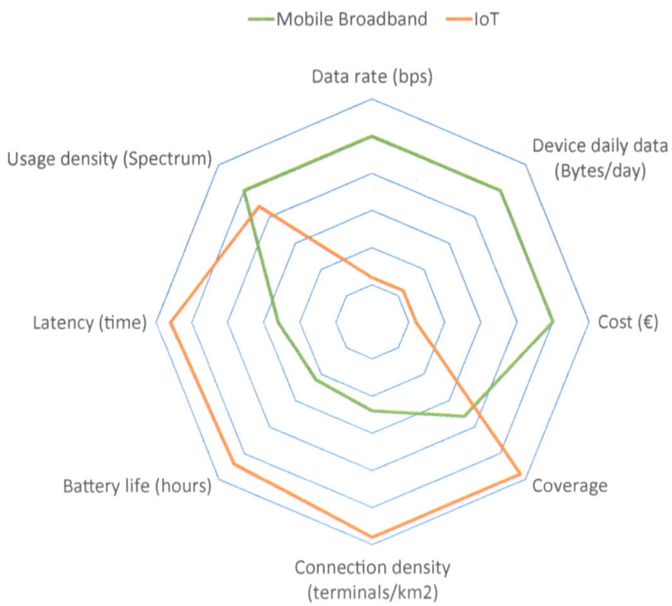

Fig. 1 Mobile Broadband traffic versus IoT traffic

Table 1 Summary of the IoT requirements and its implementation in LTE

IoT requirements	LTE implementation	Related specifications
Low cost devices	Category 0 devices	3GPP TS 36.306
	Cellular IoT	3GPP TR 45.820
Long battery life	Power saving mode	3GPP TS 23.682
	Device trigerring	3GPP TS 23.682
	Extended DRX	3GPP TR 23.770
Remote provisioning	OMA-DM	OMA device management
Flexible device identification	External identifiers	3GPP TS 23.682
Enhanced coverage	PSD boosting	3GPP 36.888
	Relaxed requirement	
	Design new channels/signals	
	Repetition	
	Low rate coding	
	Spreading	
	RS power boosting /increased RS density	
	New decoding techniques	
	Device to device communications	3GPP TR 36.843
	ProSe Relay	3GPP TR 36.814
	Heterogeneous access	3GPP TS 23.261
Managing congestion control	LAPI flag	3GPP TS 23.483
	EAB	3GPP TS 36.331
	PGW back-off timer	3GPP TS 23.401
	PGW back-off timer	3GPP TS 29.274
Managing large number of subscribers	Broadcast communications	3GPP TS 23.246
	eUICC	GSMA eUICC
Device triggering	MTC-IWF	3GPP TS 23.682
High latency communications	HLCom	3GPP TR 23.709
Architecture enhancements	For dedicated core netwtorks (DECOR)	3GPP TR 23.707
	For service capability exposure (AESE)	3GPP TR 23.708
	For cellular IoT	3GPP TR 23.720
	Group-based feature	3GPP TR 23.769
	Monitoring enhancements	3GPP TR 23.789

mobile broadband and IoT requirements. In this sense 3GPP is making a great effort to standardize LTE improvements to enhance M2M support.

In this chapter, we provide a detailed analysis of the improvements promoted by the 3GPP in order to convert LTE into a IoT capable technology and the structural changes that these modifications imply in the current LTE architecture. Table 1 provides a summary of specific LTE proposals for the improvements in the communication requirements of IoT, introduced in Sect. 2 of the chapter and the 3GPP standards developed in order to fulfil them. These standards are explained in Sects. 3 and 4.

2 Background on LTE for MTC/IoT

The 3GPP (Third Generation Partnership Project) is the standardization body that specifies LTE/LTE-Advanced, as well as 3G UTRA and 2G GSM mobile communications systems. 3GPP standards are structures as Releases. The definition of the new mobile technology known as Long Term Evolution (LTE) was initiated in 2005 and the first 3GPP specifications were introduced in Release 8 in December 2007.

The term “user equipment” (UE) have been traditionally used by 3GPP to refer to the cellular equipment utilized by the subscriber to access network services. The UE can be a smartphone or an embedded device contained in a M2M equipment. To support different UE hardware capabilities 3GPP defines different categories in 3GPP 36.306. The categories differ in the maximum data rates supported in the downlink and in the uplink, which are associated, for example, with the support of spatial multiplexing (MIMO transmission). The categories are also used by the eNodeB (LTE base station) to establish the conditions under which the communication with the UE will take place. As discussed below UE category is an important concept in IoT.

The initial version of the LTE Machine Type Communicaton (MTC) standards was realised with 3GPP Release 8 based on Category 1. Category 1 is the lowest capability with maximum bit rates of 10 Mbps downlink and 5 Mbps uplink while Category 5, the highest category in Release 8, support data rates of 300 Mbps in downlink and 75 in Mbps in uplink. Although Category 1 UEs do not support MIMO transmission, they still have 2 receiver antennas and have to support all RF bandwidth options from 1.4 to 20 MHz and other functionalities that increase the hardware complexity. Moreover this category does not meet battery, cost and rage requirements of IoT. The first release of MTC is focused on the optimization of charging mechanisms, addressing, fixed location, low mobility and low activity terminals, handling of large numbers of subscriptions and subscriber data within the network, handling issues of large number of M2M subscriptions for the user of M2M services and security aspects.

A new UE category 0 was introduced in Release 12, which achieves cost reduction of about 50% respect Category 1. Also a new power saving mode was introduced in this Release in 3GPP TS 24.301 and 23.682. The main purpose of this new feature is to reduce energy consumption while the device is not transmitting or receiving. Other solutions for MTC where discussed in Release 12 as part of 3GPP 36.888.

Improvements such as half-duplex transmission, single RF chain and reduced peak throughput are part of the specification of MTC UE Cat 0. Cost reduction using reduced system bandwidth and coverage improvements specification were postponed to Released 13.

Continuing the normative work started in Release 12 to enhance the LTE suitability for the promising IoT market, the key focus of eMTC (often referred to as LTE-M) for Release 13 is to define a new low complexity UE category type that supports reduced bandwidth, reduced transmit power, reduced support for downlink transmission modes, ultra-long battery life via power consumption reduction techniques and extended coverage operation. This is Cat-M1 (previously known as Cat-M). In Release 13 MTC achieves and additional 50% cost reduction through the restriction in system bandwidth to 1.4 MHz, which have a heavy impact on the design of the receiver, reducing the complexity of the baseband processing. This Release also introduces coverage enhancements to achieve an improvement of 15 dB, allowing operators to reach MTC devices in poor coverage conditions such as meters located in basements. These two improvements implies a main impact of the design of logical and physical channel that have to support low bandwidth modes and enhanced coverage. Another key enhancement in Release 13 is the introduction of extended DRX [4], which is based on the usage of longer DRX (Discontinuous Reception) timers to achieve further power consumption improvements. The DRX concept consists in the monitoring of downlink signalling channel during a limited period of time. During the rest of the time the device remains slept for reducing power consumption.

In Release 13 NB-CIoT and NB-LTE initiatives were started [5]. NB-CIoT is a clean slate approach promoted by Huawei. The main difference between NB-LTE and NB-CIoT comes down to how much of existing LTE networks can be reused for IoT. NB-CIoT requires new chipsets and does not seem to be backwards compatible with any LTE network older than Release 13. NB-LTE, by contrast, can be fully integrated into existing LTE networks, works within current LTE bands and does not need an overlay network. In November 2015 3GPP agreed these two initiative evolve into only one standard called Narrowband IoT (NB-IoT). NB-IoT specifies a new radio access for cellular IoT, based to a great extent on a non-backward-compatible variant of E-UTRA, that addresses improved indoor coverage, support for massive number of low throughput devices, low delay sensitivity, ultra low device cost, low device power consumption and (optimised) network architecture. So, NB-IoT has become the main steam for 3GPP LPWAN (Low Power Wide Area Network) replacing NB-LTE and NB-CIoT (Cellular IoT) proposals. On June 2016 3GPP completed the standardization of NB-IoT, the new narrowband radio technology developed for the internet-of-Things (IoT). NB-IoT UEs are referred to as Cat-NB1 (also known as Cat-M2). The reduction of their complexity, compared with Cat. 1, is up to 90%. 3GPP completed the standardization of NB-IoT on June 2016, it took only 9 months to standardize this new technology after the study phase, which demonstrates the importance of IoT for 3GPP (Table 2).

Table 2 Summary of UE categories

	Release-8	Release-8	Release-12	Release-13	Release-13
	Cat. 4	Cat. 1	Cat. 0	Cat. M1	Cat. NB1
Donwlink peak	150 Mbps	10 Mbps	1 Mbps	1 Mbps	200 kbps
Uplink peak rate	50 Mbps	5 Mbps	1 Mbps	1 Mbps	144 kbps
Number of antennas	2	2	1	1	1
Duplex mode	Full duplex	Full duplex	Half duplex	Half duplex	Half duplex
UE receive bandwidth	20 Mhz	20 MHz	20 Mhz	1.4 MHz	200 kHz
UE transmit power	23 dBm	23 dBm	23 dBm	20 dBm	23 dBm
Complexity	100 %	100 %	50 %	20 %	15 %

Table 3 Uses cases for 3GPP MTC/IoT UE categories

			Availability
LTE Cat1	Release 8	Voice services for emergency in elevators, Smart Grid Management, Kids/Elderly/Pet tracking	Available
LTE Cat0	Release 12	Cat0 is the interim solution prior Cat-M. Cat 0 is used for replacing Cat1, but cannot replace voice use cases	Available
Cat-M	Release 13	Environment monitoring, Vehicle tracking	2017
Cat-NB1	Release 13	Smart metering, smart buildings, home automation	2017

The work carried out in GSM/EDGE group inside 3GPP on Extended Coverage GSM for IoT (EC-GSM-IoT) have been joined to these two projects: LTE-M and NB-IoT.

In parallel with these standardization processes carried out inside 3GPP there are other initiatives which operate in an unlicensed spectrum basis. It is the case of LoRaWAN, Sigfox and OnRamp wireless, Weightless -N and -P etc. Most of these networks take advantage of industrial, scientific, and medical ISM unlicensed frequency bands. These technologies are currently available and have already been deployed and meet the 4 factors for LPWAN (long range, very low power, low data rate, and very low cost). Some are based on standards protocols supported by industry alliances like LoRaWAN Alliance and Weightless SIG, some are based on proprietary protocols and some are standards in-progress.

In the following sections a deeper insight into the most relevant improvements for LTE promoted by 3GPP is provided (Table 3).

3 Communication Requirements in the Internet of Things

The patterns and requirements of machine to machine (M2M) communication in the IoT are different to those of human to human (H2H) communication in traditional LTE networks. The 3GPP have identified in [6] the following requirements which are specific to M2M communications.Specifically, 3GPP TS 22.368 [1] aims to cover different market scenarios, which consist exclusively of data communications, require lower costs and efforts and service a potentially very large number of communicating terminals, which, to a large extent, generate little traffic per device.

- Addressing. The most usual pattern of communication in the IoT is expected to be many-to-one, in which many devices, sensors for example, periodically transmit data to a central server. In general, the number of devices associated with a particular server can be huge. 3GPP envisions two scenarios: the server may have an IPv6 address, in which case the devices will be assigned IPv6 addresses by the network, or the server may be reachable through an IPv4 address. In the latter case, this address may be public or private but the devices will be assigned a private IPv4 either way, due to the scarcity of public IPv4 addresses. The number of required identifiers for the devices in the network is expected to be at least two orders of magnitude higher than for H2H communications. It is therefore necessary to develop a scheme that allows the network to uniquely identify each device and the subscriber to which it belongs.
- Charging. Due to the potentially large number of devices in the network and their intermittent pattern of communication, creating detailed charging records for each device may be wasteful. The 3GPP requires the network to collect charging data with just enough granularity that it can identify the use of resources outside the limits of the subscription.
- Security. For many applications, the devices in the IoT network face security challenges which are not present in traditional H2H communications. For example, it is not uncommon for many devices to be left unattended after installation, making them potential targets for tampering, theft or destruction. In addition, depending on the scenario, the devices may be an attractive target for malicious entities which could be interested in performing, for example, a denial of service attack on them. The 3GPP requires that optimizations for M2M communications do not degrade

security with respect to H2H communications. The network should also provide secure connections between the devices and the servers.

- Triggering. For many use cases, devices in the IoT transmit data only sporadically. Sometimes, data is only transmitted when requested by a server. To optimize these use cases and avoid wasting resources, the network is required to support a device triggering mechanism. This means that the network can request a particular device to establish a connection with its server even when the device is not attached to the network or has not established a data connection.
- Low mobility. In many cases, IoT devices remain stationary for most of their lifetime. For example, power meters, pumps or vending machines remain in the same location once installed. For these devices, location updates and mobility management procedures are irrelevant and waste precious network resources. It is required that the network operator is able to change the frequency of these procedures performed by the devices.
- Time controlled communications. For some applications, it is necessary only to send or receive data at specific time intervals. As in the case of low mobility, regular procedures can become wasteful when performed outside the specified time intervals. Thus, it should be possible for the network operator to either reject or charge differently, any communication taking place outside the predefined time interval and also to alter the time interval according to local criteria such as traffic load.
- Small data transmissions. Many IoT devices need to transmit only a small amount of data at a time, typically around 1 KB. The network must support the transmission of small amounts of data, providing an optimized procedure with little overhead for this particular use case.
- Infrequent mobile terminated communications. When the device is usually the one initiating the transmissions, the network operator should be able to reduce the frequency of mobility management procedures.
- Monitoring. The network should detect events related to the status of the device, such as change of location, loss of communication and usage of the device with a different Universal Subscriber Identity Module.
- Groups. IoT devices are frequently deployed in groups. The network should provide mechanisms to handle policies for groups of devices and also to broadcast messages among the members of a group (Table 4).

Table 4 IoT communications requirements

Battery life	>10 years wiht 2AA batteries
Suscribers	Support a large number of M2M devices
Coverage	Additional 20 dB
Device cost	Very low device cost

4 Machine Type Communication (MTC)

The main objectives of eMTC (enhanced MTC) defined in Release 13 are to provide low cost devices (with prices comparable to those of GPRS/GSM devices), long battery life (around 10 years of operation with 5 Watt Hour battery), extended coverage (higher than 155.7 dB, in terms of Maximum Coupling Loss (MCL)), and variable data rates which range from 10 kbps to 1 Mbps, depending on coverage needs. The coupling loss is defined as the total long-term channel loss over the link between the UE antenna ports and the eNodeB antenna ports, and includes in practice antenna gains, path loss, shadowing, body loss, etc. The MCL is the limit value of the coupling loss at which the service can be delivered, and therefore defines the coverage of the service. The MCL is independent of the carrier frequency. It is defined in the uplink (UL) and in the downlink (DL). MTC can be deployed in any LTE spectrum and coexist with other LTE services within the same bandwidth, moreover existing LTE base stations can be reused with a software update. In the following sections these improvements will be explained in more detail together with other features introduced in previous releases.

4.1 *Low Cost LTE Devices*

In order to provide initial support for lower cost devices in LTE, 3GPP has added an additional category 0 in Release 12. The capabilities of a category 0 UE (User Equipment) are specified in 3GPP TS 36.306 [7]. The main factors enabling cost reduction in category 0 devices are those related to the reduction of the data rate by limiting the maximum transport block size (TBS) per subframe. Specially, reductions of up to 50% are envisaged.

As a reference, the maximum DL-SCH (Downlink Shared Channel) TBS has been reduced to 1000 bits from the former limit of 10296 bits in category 1 devices. Note also, the huge contrast with the highest performance capabilities such as DL Category 16 that have been also introduced in the standard in preparation for the forthcoming Gbps data rates. In addition, the number of receiving chains is also limited to a single layer.

As an optional feature, category 0 devices can also support a half-duplex type B operation. This new half-duplex scheme allows them to increase the guard period by skipping a full DL subframe before and after uplink subframes, as defined in TS 36.211 [8].

Additional specification enhancements are expected in 3GPP release 13 in the context of the proposals from TS 36.888 [9]. Multiple aspects are being analyzed to further reduce the cost of devices while enhancing coverage where required.

Specifically, up to 75% cost reductions [10] compared to category 1 are expected, mainly linked to the limitation of the reception bandwidth used and data rates, reduced transmission mode support and reduced HARQ (Hybrid Automatic Repeat Request).

4.2 Long Battery Life

Many use cases for IoT demand that devices can be deployed once and remain unattended for their lifetime. In many cases, these devices are powered by batteries that can never be recharged. A Nokia white paper [11] estimates that these devices should be able to operate for at least 10 years, drawing power from a pair of long life AA batteries.

Release 12 introduces a power saving mode (PSM) that can dramatically extend battery life for devices that send data from time to time. When a device supports PSM, the network sets the duration of an active timer during the Tracking Area Update (TAU) procedure. Once the device switches to idle mode, the timer starts. While the timer is running, the device remains in idle mode, checking for paging during Discontinuous Reception (DRX) as usual. When the timer expires, the devices enter PSM, stopping any checking for paging and becoming unreachable until the device initiates transaction. Additional improvements may be studied for Release 13. The authors of [11] note that there is a trade-off between reachability of the device and battery life extension when using longer sleep cycles. For example, when establishing the DRX cycle to 2 min, the lifespan of a device powered with two AA batteries which transmits 100 bytes daily reaches 111 months with no additional modifications.

4.2.1 Remote Management

IoT devices might be difficult to be accessed, so remote management tools are useful to guarantee the proper functionality during the lifetime of the device. The Open Mobile Alliance has defined OMA-DM (OMA Device Management) [12] that provides functionality for provisioning, configuration, software upgrades and fault management; the protocol can be transported over different wireless transports including GPRS and LTE.

4.3 Device Identification

IoT devices might be small, sealed and therefore difficult to reach once they have started operating, so the current solution of using different USIM (Universal Subscriber Identity Module) for each operator is unworkable. To overcome this situation, the GSMA has standardized [13] a remote provisioning architecture that enables the change of the data stored in the UICC (Universal Integrated Circuit Card) over the air. Using the architecture defined by the GSMA the following parameters can be provisioned:

- Network Access Credentials and Algorithms, including the security key, the Milenage parameters or any other required algorithms.

- Network Information, such as the PLMN (Public Land Mobile Network), the MSISDN (Mobile Station Integrated Services Digital Network), etc.
- Mobile Network Operator Applications, roaming management, backup services, etc.

This functionality requires a new component in the network, the Subscription Manager, which is an entity that stores the eUICC access keys, which is trusted by one or more Mobile Network Operators. Apart from this information, IoT devices may have external identifiers, which are unique identifiers that can be used by external applications, servers to communicate with the UE from outside the network without knowing its IP (see section Device Triggering). These identifiers should have a domain identifier component, used by the mobile network operator to identify the services that can be accessed for this UE and the local identifier, which must permit the derivation of the IMSI (International Mobile Subscriber Identity).

4.4 Enhanced Coverage

TS 36.888 [9] introduces several techniques to obtain 20 dB of improvement in the coverage. A first set of techniques is composed by TTI (Transmission Time Interval) bundling, HARQ retransmission, repetition, code spreading, RLC (Radio Link Control) segmentation, low rate coding, low modulation order and new decoding techniques, which can be used to accumulated energy in order to improve coverage by prolonging the transmission time. Power boosting can be also used to transmit more power from the base station or Evolved Node B (eNB) to the MTC UE, while PSD (Power Spectral Density) boosting enables power to be concentrated in a reduced bandwidth at the eNB or the UE. Due to relaxed requirements of MTC communications, performance of some LTE channels can be relaxed. In the case the coverage improvement requirement cannot be fulfilled, new channels or signals can be designed. Also, small cell deployments can be used to improve coverage. Which techniques depends on the channel, as shown in Table 5.

Existing coverage improvement solutions deployed for normal LTE UE such as directional antennas, external antennas can be applied. Single Frequency Network (SFN) multicast may also be used. This technology consists in sending redundant broadcast signals from all cells, helping to increase coverage especially at the cell edges. Finally, LTE Direct device-to-device (D2D) communications [14] can help extend coverage. D2D was introduced in Release 12 for traditional subscribers and public safety use cases enabling reliable one-to-many communications between devices in and out of coverage. In Release 13, D2D has been extended with new discovery and communication mechanisms such as out-of-coverage and multi-carrier. Also device-to-network relays [15] have been introduced for Public Safety uses cases. For Release 14 and beyond, additional D2D communications capabilities are being considered such as multi-hop communication, which has been proposed for vehicle-to-vehicle (V2V) communications.

Table 5 Possible link-level solutions for coverage enhancement of physical channels and signals [3GPP TR 36.888]

Channel/Signals solutions	PSS/SSS	PBCH	PRACH	(E)PDCCH	PDSCH/ PUSCH	PUCCH
PSD boosting	x	x	x	x	x	
Relaxed requirement	x		x			
Design new channels/signals	x	x	x	x	x	
Repetition		x	x	x	x	x
Low rate coding		x		x	x	x
TTI bundling/ Retransmission					x	
Spreading		x			x	
RS power boosting/increased RS density		x		x	x	
New decoding techniques		x				

4.5 Device Triggering

The device triggering function was introduced in Release 11 (3GPP TS 23.682 [16]) in order to enable the reachability of dormant devices by MTC servers applications using a unique external identifier. With these external identifiers, the provider can access the terminals using SMS without having to allocate an IP. Once the UE receives the SMS it must be able to identify the application and trigger the appropriate actions, analysing the payload.

4.6 Managing Congestion Control

One of the most important characteristics of IoT is the expected number of connected devices connecting to the network, which can lead to congestion that can affect all users (both things and humans) in the network. This congestion could be for many reasons, simultaneous synchronization of the things, faulty transmissions or signalling storms after a period of outage in the network. LTE already provides some generic functions to avoid congestion [17], which are depicted in Fig. 4.

The eNB can detect insufficient resources and reject RRC (Radio Resource Control) connection attempts or bearer creation or reconfiguration; it also can use an RRC back-off function. The MME (Mobility Management Entity) can also reject connections, if it has received a back-off indication from the PGW (Packet Data Network Gateway). If this indication is not received, it tries to create a session with the SGW (Serving Gateway), which is forwarded to the PGW, which, in turn, can detect congestion for a specific APN (Access Point Name). If the APN is saturated the SGW can reject the session creation, setting the flag to APN congestion (3GPP TS 29.274 [18]) or it can reject a PDN (Packet Data Network) creation. This enables a back-off time for the MME which rejects all requests for a specified time (3GPP TS 23.401 [19]). Furthermore, the 3GPP has studied on solutions for managing core network overload (3GPP TR 23.843 [20]) which provide information on the use or modification of diameter, GTP-C (GPRS Tunnelling Protocol user plane Control) and SS7 interfaces to manage overload.

Another important functionality, introduced in Release 11, is the use of the Low Access Priority Indicator, a flag that can be enabled in the UEs in production or via the OMA-DM (OMA Device Management) or OTA (Over-the-air) interfaces. This flag is provided in the attach request to the eNB and is shared with the other elements of the network. In the event of congestion the flag will be used to differentiate priority from non-priority traffic and also to increase the timer to trigger location update procedures, reducing the overload for this type of equipment. The terminals using this flag can also send messages with normal priority (for instance for terminals that might require higher priority in certain situations like alarms or emergencies). To reduce signalling procedures, the Extended Access Barring (3GPP TS 36.331 [21]) was introduced in Release 10. Using this feature the cell can indicate, via SIB14 (System Information Block), that the access is barred for certain access classes so no connection attach is attempted.

4.7 Heterogeneous Access

Support for seamless offloading to heterogeneous radio access might be useful when deploying a MTC architecture as it can provide coverage in zones without LTE. 3GPP has specified the access for GSM, UMTS, HSPA, etc. and now has included network elements to support seamless offloading to Wi-Fi (3GPP TS 23.261 [22], 3GPP TS 23.402 [23]). The functionality provides seamless handover which might not be very useful for IoT applications but provides trusted WLAN access, which might improve the security and enable the use of MTC application servers deployed in 3GPP networks.

4.8 Security Considerations

Devices in the IoT can be a potential target for malicious attacks due to, among other factors, their potentially limited processing and memory resources, the monetary or strategic value of the device to which they are attached, or the potential to launch denial of service attacks due to the large number of devices present in the network. Legacy 2G systems are considered insecure and there are already a number of vulnerabilities that have been discovered in IoT systems. Cellular IoT systems should be designed under the assumption that communications can be eavesdropped and thus, additional layers of encryption should be used [24].

As mentioned, the large number of devices that the IoT can serve makes the possibility of signalling storms more likely. Specifically, NAS (Non-Access Stratum) signalling procedures have been identified as a potential target for this. Another possible attack of this type can be performed by sending small data packets to a large number of devices, creating a huge number of RRC state transitions and potentially overloading the network core. Even if not created by a malicious entity, signalling storms can occur spontaneously due to the nature of the pattern of communication that many IoT devices exhibit, consisting in small but frequent transmissions. As in the previous case, these bursts translate into a large number of RRC state transitions, leading to congestion or denial of service. This is especially likely when many devices try to transmit data at the same time which can happen, for example, when these devices come back online after a power outage. The 3GPP is actively working on techniques to avoid the likelihood of such signalling storms.

4.9 MTC Improvements in the Core Network

3GPP 23682 [16] specifies architecture enhancements to improve MTC communications according to the use cases and service requirements defined in TS 22.368 [6], TS 22.101 [25], and related 3GPP requirements specifications.

Figure 2 provides the MTC architecture with the new elements introduced to provide access to network configuration to third party applications as well as functionalities like device triggering.

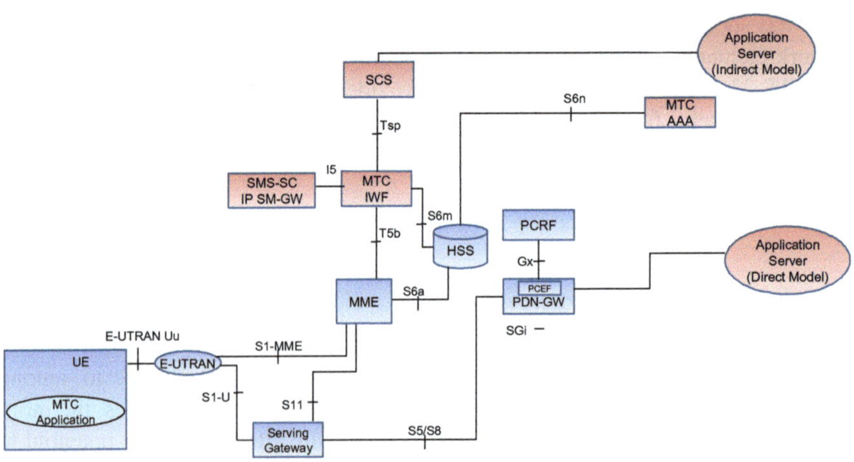

Fig. 2 LTE Architecture for MTC

The MTC architecture enables the communication between a UE running an MTC application and MTC services, as well as optionally providing some network functionality via the SCS (Service Capability Server). The components of the architecture are:

- MTC AAA (MTC Authentication, Authorization and Accounting), is in charge of returning external identifiers associated with an IMSI, and it might query the HSS to retrieve the values. The component can act as a proxy, in that case it will translate from IMSI to external identifier and vice versa.
- MTC IWF (MTC Inter Working Function), in charge of displaying device trigger functionalities for the SCS (reception of request, use of identifiers to activate the UE, selection of the mechanism, etc.), authenticate connection request from the SCS.
- SCS, an entity that connects to the 3GPP network exposing functionality that can be used by one or more MTC applications. The model of deployment can be direct, when the application server directly accesses the operator network, or indirect, when the application server accesses to the functionality via the SCS (which can be owned by the operator or by the MTC service provider), or hybrid, a mixture of both. This element is directly related to the device triggering function, introduced in the previous section. The SCS will send an SMS that contains the Trigger Payload, which is the information destined for the application running in the UE and the information to route it. The SCS will also provide functions to replace a previously sent SMS with a new one.

The network exposes functionality to the MTC applications or the SCSs via standard interfaces. For instance, the SCS and the AS (Application Server) could access the establishment of dedicated bearers via the Rx interface, the device triggering function using the Tsp interface, content delivery via the BM-SC or any other function exposed to third parties in the 3GPP specifications. Figure 3 depicts some of the elements that might be accessed via third party APIs based on the SCEF (Service Capability Exposure Function), a new node introduced specifically for machine type data. This element is part of the Architecture Enhancements for Service Capability Exposure (AESE) [26] introduced in Release 13, which displays network services to 3rd parties.

Other architecture enhancements introduced in Release 13 and centred on M2M communications are: Monitoring Enhancements (MONTE) [27], which shows network information to 3rd parties for troubleshooting and Group based Enhancements (GROUPE) [28], which covers group based policies and group based addressing, Dedicated Core Networks (DECOR) [29], which enables core network nodes be selected based on subscription information (for example, special MME for M2M users). Finally optimizations to support high latency communications (HLCom) [30] are also relevant items for IoT. An introduction to these enhancements is provided in [31].

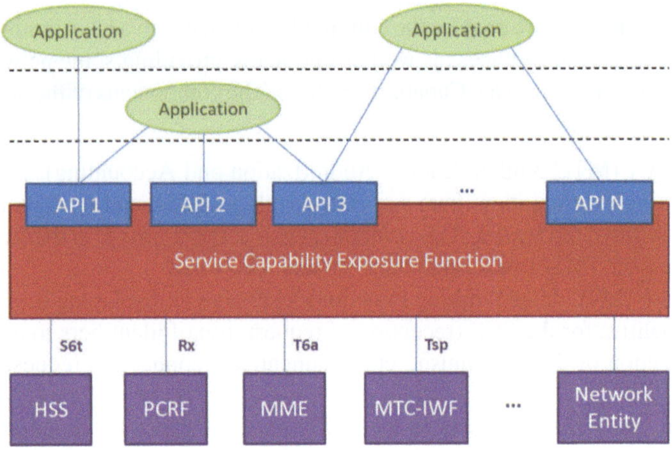

Fig. 3 Service capability exposure function

5 Narrowband IoT (NB-IoT)

As result of the 3GPP study item TR 45.820 [5] a new technology have been proposed, called NB-IoT. NB-IoT is a new narrowband radio technology being standardized in 3GPP to address the requirements of the Internet of Things (IoT). NB-IoT is being integrated in the LTE standard (3GPP TS 36.212, 3GPP TS 36.102, 3GPP TS 36.321, 3GPP TS 36.331, etc.). A new UE category has been standardized in 3GPP TS 36.101, Cat-NB1. This new technology will provide improved indoor coverage, support of massive number of low throughput devices, low delay sensitivity, ultra-low device cost, low device power consumption and optimized network architecture.

Unlike eMTC that can only be deployed in-band, utilizing resource blocks within a normal LTE carrier, NB-IoT can also be deployed in the unused resource blocks within a LTE carrieŕs guard-band, or standalone for deployments in dedicated spectrum. NB-IoT is also particularly suitable for the re-farming of GSM channels.

The requirements of NB-IoT are the same that for MTC, but with the focus on the low-end massive MTC scenarios. NB-IoT provides even lower cost in UEs than eMTC and a higher extended coverage of 164 dB MCL (at least for standalone deployment). The battery life is also around 10 years with 5 Watt Hour battery. NB-IoT is compatible with the enhanced DRX mode introduced in Release 13 to reduce power consumption. The number of devices supported per cell is around 50.000.

5.1 *Core Network for IoT*

NB-IoT architecture simplifies the existing EPC architecture in order to address IoT requirements. It bases small data transmission on a simplified architecture that is oriented at transporting small data over NAS signalling messages. A dedicated core network node for CIoT profile provides combined C-plane and U-plane functions, e.g. aggregating some of the functions traditionally reside in MME and SGW (and in some instance PGW) in a new logical entity called C-SGN. C-SGN can be implemented to support only the necessary functionality required for CIoT use cases. S1-lite is an optimized version of S1-C (the S1 control plane which is based in the S1AP protocol) interface between the eNodeB and the MME. From S1-C only the necessary S1AP messages are supported (and inside the messages only the required parts) for the related CIoT procedures and the optimised security procedures are supported, as well as some functionality to support data transport. The user plane data is carried in these modified S1AP messages in order to support efficient small data handing. The data planes tunnels of the S1-U interface (defined in TS 23.401) are not required.

This architecture supports certain modifications (e.g. native support of SMS in PS domain, attach without PDN connection, simplified NAS etc.) to be applicable only to CIoT UEs without requiring procedures to maintain backwards compatibility with other UEs.

In NB-IoT, for signalling optimization, two options are introduced in addition to legacy RRC connection setup, mandatory Data-over-NAS (Control Plane optimization, solution 2 in [32]) and optional RRC Suspend/Resume (User Plane optimization, solution 18 in [32]). In the control plane optimization the data is sent over Non Access Stratum (NAS), directly from the Mobility Management Entity (MME) in the core network to the UE without interaction from the base station.

The RRC Suspend/Resume procedures reduce the signalling overhead and also improve UE battery lifetime. The protocol architecture of NB-IoT and LTE is separated into control plane and user plane. Control plane consists of protocols which control the radio access bearers and the connection between the UE and the network. The highest layer of control plane is called Non-Access Stratum (NAS), which conveys the radio signalling between the UE and the EPC, passing transparently through the radio network. It is responsible for authentication, security control, mobility management and bearer management. Access Stratum (AS) is the functional layer below NAS, and in control plane it consists of Radio Resource Control protocol (RRC). RRC configures the user and control planes according to the network status. There are two main RRC states, RRC_Idle or RRC_Connected, and the RRC entity controls the switching between these states. In RRC_Idle, the network knows that the UE is present in the network and the UE can be reached in case of incoming call. In RRC_Connected the UE has a connection to the eNodeB, the network knows the UE location on cell level and the UE may receive and transmit data. Connection is released, switching to RRC_Idle, when there is no traffic to save the UE battery and radio resources. RRC Suspend/Resume procedures reduce the signalling overhead required for UE state transition from Idle to Connected mode in order to have a user plane transaction with the network and back to Idle state by reducing the signalling messages required compared to legacy operation (Fig. 4).

In order to support RRC Suspend/Resume procedures MME, eNodeB and UE have to include new functionalities. The Mobility Management Entity (MME), which is responsible for handling the mobility of the UE, has to support the new procedures on

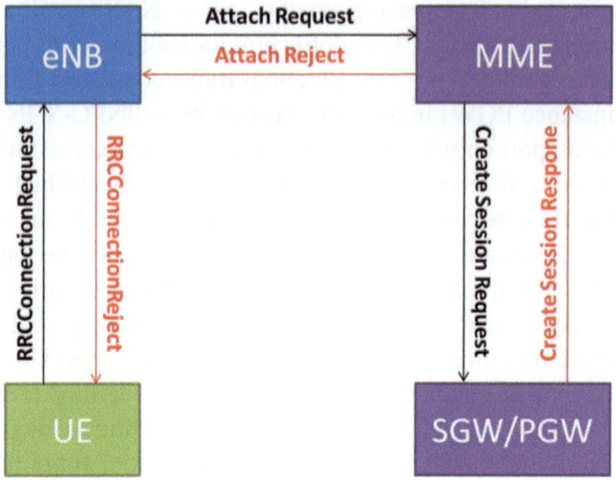

Fig. 4 Congestion detection in the different elements of the network

S1AP, in particular the S1AP procedure triggered by the "RRC Suspend" and "RRC Resume" procedures on the radio interface. The MME has also to store the UE Context information when the RRC connection is suspended and the S1AP association is kept. The eNodeB should include support for storing the UE context as well as the S1AP association related parameters when the UE is in the new CIoT RRC-Idle state. Moreover, the eNodeB has to support the new procedures to suspend/resume the connection. These procedures will be implemented between the eNB and the UE as well as between the eNB and the MME. Finally the UE has to store relevant AS information when the UE enters into the CIoT RRC_Idle and support the procedures needed to suspend and resume the RRC connection.

Data-over-NAS is considered as the basis for the normative work for support of infrequent small data transmission (for IP data, non-IP data and SMS). Support of solution 2 is mandatory for both the UE and the network. This solution provides efficient support of infrequent small data transmission for IoT and support for non-IP data. The solution is based on lightweight core network architecture depicted in Fig. 5. This solution piggybacks uplink small data onto the initial NAS uplink message that is extended and it uses an additional NAS message for carrying downlink response small data. The efforts for setting up the user plane, i.e. related RRC messages and AS security setup, can be avoided. There is not establishment of DRBs and S1-U bearers. Data can be also delivered through the SCEF, which is used for the delivery of non-IP data over control plane. The implementation of this solution also implies modifications of the procedures available at the UE, eNodeB and C-SGN/MME to support i.e. handling of "NAS small data" sessions and the attach without bearer activation.

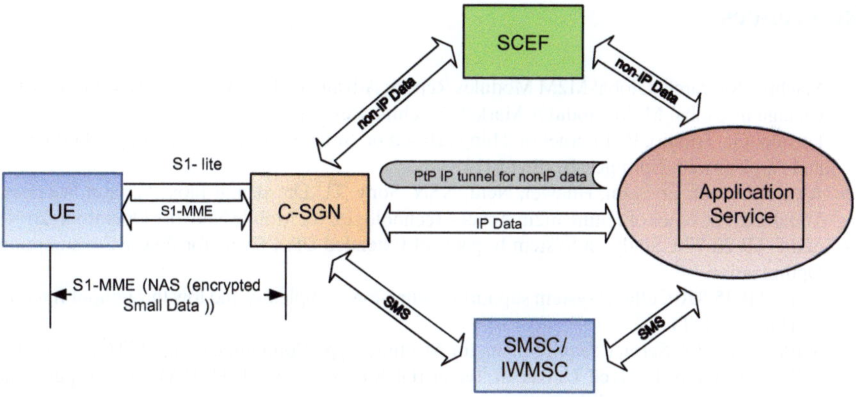

Fig. 5 LTE Architecture for NB-IoT

6 Conclusion

In this chapter two different LTE approaches for the delivery of IoT communications have been introduced, LTE-M and NB IoT. In 3GPP Release 12 and Release 13 LTE-M introduce optimizations for efficient machine-type communications. Concretely, the improvements in Release 12 have reduced by 50%, the complexity of UEs compared to Category 1, and have extended the battery life to 10 years for downlink delay-tolerant traffic. Release 13 has introduced techniques to increase coverage by 15–20 dB, the UE complexity has been reduced by a 75% and battery life has been extended beyond 10 years for new use cases. Furthermore these optimizations can be combined with other existing 3GPP technologies providing an excellent solution for IoT scenarios.

The improvements presented in this chapter shown the evolution of LTE technology to tackle IoT requirements, providing multi-year battery life, reduced complexity, deeper coverage and high node density. These enhancements together with the advantage offered by standard LTE (flat and flexible architecture, built-in security, spectral efficiency, etc.) make this technology the future of IoT.

While LTE-M is a pure LTE technology NB-IoT is a new radio technology that coexists with existing GSM, UMTS and LTE systems.

In regard to the uses cases of both technologies the main differences are mobility and throughput. NB-IoT offers reduced data rate and mobility support compared to LTE-M.

In Release 14, we can expect further enhancements for NB-IoT in the form of multicast support and latency reduction optimizations to enable vehicle to vehicle (V2V) communications.

Acknowledgements This work has been funded by the Spanish Ministry of Economy and Competitiveness (grant TIN2015-67083-R) and FEDER.

References

1. Machina Research. Global M2M Modules Report: Advancing LTE Migration Heralds Massive Change in Global M2M Modules Market. Machina Research (2013)
2. Fortino, G., Trunfio, P.: Internet of Things Based on Smart Objects. Technology, Middleware and Applications. Springer, Berlin (2014)
3. Alcatel Lucent, Ericsson, Huawei, Neul, NSN, Sony, TU Dresden, u-blox, Verizon Wireless, Vodafone, A Choice of Future m2m Access Technologies for Mobile Network Operators (2014)
4. 3GPP TR 23.770, Study on System Impacts of Extended DRX Cycle for Power Consumption Optimization
5. 3GPP TR 45.820 Cellular system support for ultra-low complexity and low throughput Internet of Things (CIoT)
6. 3GPP TS 22368, Service requirements for Machine-Type Communications (MTC); Stage 1
7. 3GPP TS 36306, Evolved Universal Terrestrial Radio Access (E-UTRA); User Equipment (UE) radio access capabilities
8. 3GPP 36.211, Evolved Universal Terrestrial Radio Access (E-UTRA); Physical channels and modulation
9. 3GPP 36.888, Study on provision of low-cost Machine-Type Communications (MTC) UEs based on LTE
10. Ericsson and Nokia Siemen Networks, LTE Evolution for Cellular IoT (2014)
11. Nokia White Paper, Optimizing LTE for the Internet of Things (2014)
12. Open Mobile Alliance, Device Management Architecture (2013)
13. GSMA Association, Embedded SIM Remote Provisioning Architecture, Version 1.1 (2013)
14. 3GPP TR 36,843, Study on LTE device to device proximity services; Radio aspects
15. 3GPP TR 36.814, Evolved Universal Terrestrial Radio Access (E-UTRA); Further advancements for E-UTRA physical layer aspects
16. 3GPP TS 23.682, Architecture enhancements to facilitate communications with packet data networks and applications
17. NTT Docomo Technical Report, Core Network Infrastructure and Congestion Control Technology for M2M Communications (2013)
18. 3GPP TS 29.274, 3GPP Evolved Packet System (EPS); Evolved General Packet Radio Service (GPRS) Tunnelling Protocol for Control plane (GTPv2-C); Stage 3
19. 3GPP TS 23.401, General Packet Radio Service (GPRS) enhancements for Evolved Universal Terrestrial Radio Access Network (E-UTRAN) access
20. 3GPP TR 23.843, Study on Core Network Overload (CNO) solutions
21. 3GPP TS 36.331 Evolved Universal Terrestrial Radio Access (E-UTRA); Radio Resource Control (RRC); Protocol specification
22. 3GPP TS 23.261, IP flow mobility and seamless Wireless Local Area Network (WLAN) offload; Stage 2
23. 3GPP TS 23.402, Architecture enhancements for non-3GPP accesses
24. Piqueras Jover, R.: Security and Privacy in the Internet of Things (IoT): Models, Algorithms, and Implementations (2015)
25. 3GPP TS 22.101 Service aspects; Service Principles
26. 3GPP TR 23.708 Architecture Enhancements for Service Capability Exposure (AESE)
27. 3GPP TR 23.789 Monitoring Enhancements (MONTE)
28. 3GPP TR 23.769 Technical Specification Group Services and System Aspects; Group based Enhancements (GROUPE)
29. 3GPP TR 23.707 Architecture enhancements for dedicated core networks; Stage 2
30. 3GPP TR 23.709, Study on optimizations to support high latency communications; Stage 2
31. Kunz, A., Prasad, A., Samdanis, K., Husain, S., Song, J.: Enhanced 3GPP system for machine type communications and internet of things. In: 2015 IEEE Conference on Standards for Communications and Networking (CSCN), pp. 48–53. Tokyo (2015)
32. 3GPP TR 23.720 Study on architecture enhancements for Cellular Internet of Things

Towards Efficient Deployment in Internet of Robotic Things

Cristanel Razafimandimby, Valeria Loscri and Anna Maria Vegni

Abstract Internet of Robotic Things (IoRT) is a new concept introduced for the first time by ABI Research. Unlike the Internet of Things (IoT), IoRT provides a dynamic actuation and is considered as the new evolution of IoT. This new concept will bring new opportunities and challenges, while providing new business ideas for IoT and robotics' entrepreneurs. In this work, we will focus particularly on two issues: (i) connectivity maintenance among multiple IoRT robots, and (ii) their collective coverage. We will propose (i) IoT-based, and (ii) a neural network control scheme to efficiently maintain the global connectivity among multiple mobile robots to a desired quality-of-service (QoS) level. The proposed approaches will try to find a trade-off between collective coverage and communication quality. The IoT-based approach is based on the computation of the algebraic connectivity and the use of virtual force algorithm. The neural network controller, in turn, is completely distributed and mimics perfectly the IoT-based approach. Results show that our approaches are efficient, in terms of convergence time, connectivity, and energy consumption.

Keywords IoT-based · Connectivity maintenance · IoRT · Neural network

1 Introduction

Nowadays, Internet of Things (IoT) technology begins to take an important place in economic systems and in society daily life [5, 6, 12, 15, 16]. It has got a large success in several application areas, ranging from smart city applications [4, 24] to smart grid [25]. However, most of IoT applications are based only on static actuation. Later,

C. Razafimandimby (✉) · V. Loscri
Inria Lille – Nord Europe, Villeneuve, France
e-mail: jean.razafimandimby_anjalalaina@inria.fr

V. Loscri
e-mail: valeria.loscri@inria.fr

A.M. Vegni
University of Roma Tre, Rome, Italy
e-mail: annamaria.vegni@uniroma3.it

R. Gravina et al. (eds.), *Integration, Interconnection, and Interoperability of IoT Systems*, Internet of Things, DOI 10.1007/978-3-319-61300-0_2

adding an active role for actuators will be needed, in order to optimize the systems where they are present. Robotic systems match very well to this new need, since robots can sense and interact with their environment. Therefore, integrate robots as a device in IoT is obvious.

1.1 Overview of Internet of Robotic Things

Internet of Robotic Things (IoRT) is a new concept introduced for the first time by ABI Research [1]. It was defined as an intelligent set of devices that can monitor events, fuse sensor data from a variety of sources, use local and distributed intelligence to determine a best course of action, and then act to control or manipulate objects in the physical world [1]. This new concept is expected to be the evolution of IoT and robotics (Fig. 1).

In most of the time, the brain of robots and control mechanisms are local i.e. on-board the robots themselves. However, in IoRT concept, the computation and the control can be assigned to the cloud. In this case, Internet allows the IoRT robots to be connected to the Cloud. With advances in robot operating system framework, communication with the Internet is not complicated. It requires only a simple call of an application programming interface (API).

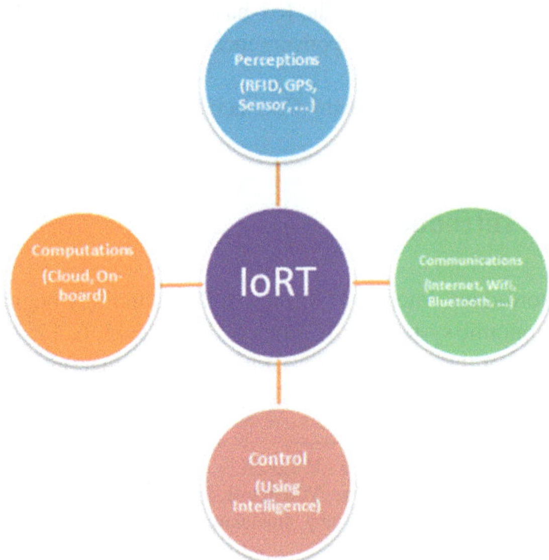

Fig. 1 Internet of Robotic Things

Assigning the processing to the Cloud increases the data-processing and interpretation capabilities. However, even if Cloud computing plays important role in the IoRT concept, local and distributed computations are still required for real-time applications or when the access to the Cloud is not feasible (e.g. after a disaster).

As mentioned before, local and distributed intelligence are also used to determine a best course of action. Hence, intelligence is also a vital part for the IoRT concept. The differents kinds of intelligences used in this work will be detailed in Sects. 3 and 4.

1.2 Issues and Motivations

In various IoRT applications (e.g.: smart agriculture, smart environment monitoring, smart exploration, smart disaster rescue, etc.), the use of mobile robots' teams brings many advantages over one powerful IoRT robot. As a matter of fact, a team of robots can accomplish tasks more efficiently, faster and more reliable than a single robot [7, 9, 22]. To carry out cooperative tasks, IoRT team members need to communicate with each other, often via a wireless link (i.e. Wifi, Bluetooth). Maintaining communication among multiple mobile IoRT robots is therefore a crucial issue.

Many approaches have been designed to maintain the connectivity of multi-robot and multi-agent systems. These approaches can be classified into two groups i.e. (i) local and (ii) global connectivity maintenance. With the local connectivity maintenance, the initial set of edges which define the graph connectivity must be always preserved over time. Unlike local connectivity maintenance, global connectivity maintenance allows suppression and creation of some edges, as long as the overall connectivity of the graph is conserved. Different examples and results related to these approaches can be found in [8, 10, 18, 19, 27, 28].

In Multi-Robot Systems, global connectivity maintenance is often used since the local connectivity maintenance presents some restrictions. Besides connectivity maintenance, the major problematic in global connectivity maintenance approaches is how to maximize the network connectivity. Maximizing the connectivity is important to ensure reliable communication between any pair of IoRT robots. Many works based on graph theory were proposed in the literature to face this problematic. These works are extensively used in multi-agent systems and are based on the maximization of the algebraic connectivity. In this work, we try to migrate some of these ideas in IoRT applications. Precisely, we will use the graph connectivity metric to maintain the global connectivity of IoRT robots' team, when they are in mobility.

This work addresses also the coverage issue. In general, coverage issue aims to determine how well the sensing field is monitored or tracked by sensors. In literature, Virtual Force Algorithm (VFA) was widely used to formulate this problem [31]. However, these methods have limitations since there are situations that do not allow the systems to converge in a stable state [11]. We will present a new solution to this problem later in this work.

To summarize, in this work we address two problems i.e. (i) connectivity maintenance, and (ii) collective coverage. However, it is interesting to mention that maximizing these two parameters simultaneously is difficult (if not impossible). Maximize the collective coverage may lead poor communication quality and conversely (i.e. a very good communication may lead poor coverage). It follows that our goal is to capture the trade-off between collective coverage and communication quality. Leveraging on the above motivations, in this work, we propose two motion control strategies which maintain global connectivity between IoRT robots to a desired QoS level. The first approach is an IoT-based while the second is a distributed trained neural network controller.

The main contribution of this work is the design of two approaches that have the following properties:

- The first approach is IoT-based, IoRT robots works with a central object which has high computation capability for network connectivity computing/monitoring and for the robot motion decision;
- Our both approaches converge to the desired communication quality level;
- Connectivity between any pair of IoRT robots is kept all along the deployment procedure;
- Our approaches use a distributed virtual force algorithm when the access to the central object is available and a distributed trained neural network controller otherwise. Both strategies are computed locally and based only on the local neighborhood information.

The rest of this chapter is organized as follows. Section 2 provides some backgrounds which include information on algebraic graph theory and neural networks. Section 3 describes the IoT-based approach, while Sect. 4 details the neural network approach. Section 5 provides the simulation results. Finally, Sect. 6 is dedicated to the conclusions.

2 Backgrounds

2.1 Graph Representation and Eigenvalues

Multi-Robot Systems (MRS) can be represented by a graph $G(V, E)$ where V is the set of vertices representing each IoRT robot and $E \subseteq V^2$ is the set of edges. E can be defined as: $E = \{(i, j) \in V^2 \mid i \neq j \wedge d(i, j) \leq R\}$, where $d(i, j)$ is the euclidean distance between i-th and j-th IoRT robots and R is the communication range. Following the above definition, let N_i be the one-hop neighborhood of the i-th IoRT robot. Thus, N_i is the set of IoRT robots which can exchange information with IoRT robot i. N_i can be defined as follow:

$$N_i = \{j \in V \mid d(i, j) \leq R\} \tag{1}$$

2.1.1 Definition 1

An undirected graph G is connected if there exists a path between each pair of vertices.

In this paper, the graph G may evolve over time due to the IoRT robots motion but has to be always connected.

2.1.2 Definition 2

Let define a matrix $A \in R^{n \times n}$. The scalar λ is an eigenvalue of A if there exists a non-zero vector w such that

$$A.w = \lambda.w \tag{2}$$

The vector w is called eigenvector of A corresponding to λ.

2.2 *Laplacian Matrix and Algebraic Connectivity*

Given a undirected graph G, its Laplacian matrix L is defined as:

$$L(G) = \Psi(G) - A(G) \tag{3}$$

where:

- $A(G)$ is the weighted adjacency matrix of graph G whose entries A_{ij} is defined as in [18]:

$$A_{ij} = \begin{cases} 1 & d(i,j) < D_{th} \\ e^{\frac{-5(d(i,j)-D_{th})}{R-D_{th}}} & D_{th} \leq d(i,j) \leq R \\ 0 & d(i,j) > R \end{cases} \tag{4}$$

- $\Psi(G)$ is a diagonal matrix such as the components $\Psi_i = \sum_{i=1}^{n} A_{ij}$ along the diagonal
- D_{th} is the desired distance between each pair of IoRT robots.

The Laplacian matrix $L(G)$ holds some interesting properties:

1. Let **1** be the column vector of all ones. Then, L**1** = 0.
2. Let λ_i, $i = 1, \ldots, n$ the eigenvalues of the Laplacian matrix $L(G)$.

- The eigenvalues of $L(G)$ can be ordered such that

$$0 = \lambda_1 \leq \lambda_2 \leq \lambda_3 \leq \cdots \leq \lambda_n \tag{5}$$

- $\lambda_2 > 0$ if and if only the graph G is connected. The second-smallest eigenvalue λ_2 is called also algebraic connectivity of the graph G. The value of λ_2 indicates how weel connected the graph is.

Further details on graph theory and proof can be found in [13].

2.3 Artificial Neural Networks

Artificial Neural Networks (ANN) were inspired by the human brain and was designed as a computational model to solve specific problems [2]. It's architecture is defined by (i) a basic processing element called artificial neuron, and (ii) the way in which they are interconnected. The output value of a neuron is given by:

$$output = f(\sum_i w_i x_i + b) = f(W^T X + b) \tag{6}$$

where

- x_i: the inputs
- w_i: connections' weights between x_i and the neuron
- W: weights' vector
- X: inputs' vector
- b: the bias
- f: the activation function

The basic architecture of ANN contains three neuron layers: input layer, hidden layer and output layer. In this case, the outputs of one layer become the inputs of next layer [2]. A typical artificial neuron and a basic ANN are illustrated in Fig. 2.

A key element of an artificial neural network is its ability to learn. This meant that ANN has to learn from a data set in order to match the inputs to the desired output. During the learning process, weights and biases are adjusted till the desired output

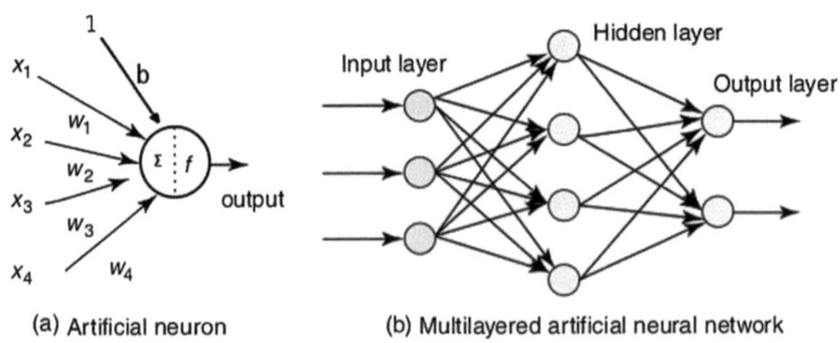

Fig. 2 Architecture of an artificial neuron and a multilayered neural network

will be reached. There are several learning algorithm but in this work we will use the backpropagation algorithm [26].

2.4 Virtual Force Algoritm

Virtual Force Algorithm (VFA) is extensively used to solve the coverage problem on robots and sensors networks. The main idea is to model each robot or sensor as a particle in the potential field. The potential field exerts forces on the nodes nearby. The force may be either attractive or repulsive force according whether they are close or far to each other. If two nodes are placed closer than the desired distance D_{th}, repulsive forces are exerted on each other. Otherwise, attractive forces are exerted if two nodes are farther than D_{th}. The repulsive force aims to avoid a poor coverage while the attractive force ensures that a globally uniform node placement will be achieved [31]. For any pairwise of node i and j, the mutual force F_{ij} can be written as the negative gradient of the potential field. So, we can build a potential function V_{ij} such as:

$$F_{ij} = -\nabla V_{ij} \tag{7}$$

According to the traditional VFA, the force $\overrightarrow{F_{ij}}$ is given as:

$$\overrightarrow{F_{ij}} = \begin{cases} (w_a(d(i,j) - D_{th}), \theta_{ij}), & if\ d(i,j) > D_{th} \\ 0, & if\ d(i,j) = D_{th} \\ (w_r(d(i,j) - D_{th}), \theta_{ij} + \pi), & if\ d(i,j) < D_{th} \end{cases} \tag{8}$$

where:

- w_a is the virtual force attractive coefficient
- w_r is the virtual force repulsive coefficient
- θ_{ij} is the orientation of the line segment from nodes
- D_{th} is the desired distance between each pair of nodes
- $d(i, j)$ is the euclidean distance between nodes i and j

This traditional VFA has limitations since there are situations that do not allow the systems to converge in a stable state [11]. We will present a new modified version of VFA in Sect. 3.

2.5 Ideal Deployment for Full Coverage

The problem of maximizing the coverage of robots and sensors network was addressed in several works, using either virtual force algorithm [17, 21, 30, 31] or geometrical approaches [3, 20, 29] which manage the pairwise distance between

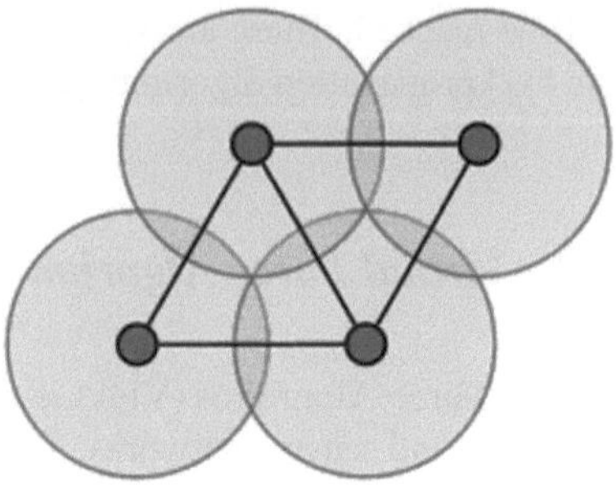

Fig. 3 Ideal deployment for full coverage

any pair of node and locally arrange the network topology as a triangle tessellation. In this work, we focus only on the virtual force algorithm, since geometry-based approaches computation can only be done when the global location information of all the nodes in the network is known. An ideal deployment structure is show in Fig. 3. It is worth to mention that the ideal deployment will be achieved if the proposed approach converges to the desired distance D_{th}. There is no coverage hole in an ideal deployment.

3 IoT-Based Approach

As we saw in the previous Section, an undirected graph is connected if and only if $\lambda_2 > 0$. Therefore, any strategy which maintains λ_2 at positive values guarantees global connectivity among multi-robot systems.

In this Section, we present an IoT-based approach which is capable of maintaining desired wireless communication coverage among neighboring robots.

The proposed approach uses a Central Object (CO) with high computation capability to compute and monitor the connectivity of the overall multi-robot system. We assume that each IoRT robot knows its own position by using GPS or other localisation system. Beacon messages are used to allow IoRT robots exchange their positions with their one-hop neighbors. Each IoRT robot in the system applies a modified version of VFA (as described on formula 9) to control its movement. This computation is only based on the local neighborhood information. In order to keep the desired distance and hence the desired connectivity quality with its neighbor, the i-th IoRT robot should move away from the IoRT robot $j \in N_i$ if $d(i, j) < D_{th}$ and should move close if $d(i, j) > D_{th}$. D_{th} is the desired distance between each pair of IoRT robots. This simple control law generates a vector position $\overrightarrow{P_{ij}}$ such that the i-th IoRT robot keeps the line of sight of the IoRT robot j. $\overrightarrow{P_{ij}}$ is defined as:

$$\overrightarrow{P_{ij}} = \begin{cases} (0.1 \times k \times \Delta d, \theta_{ij}) & if\ d(i, j) > D_{th}\ and\ \Delta d > \epsilon \\ (k \times \Delta d, \theta_{ji}) & if\ d(i, j) < D_{th}\ and\ \Delta d > \epsilon \end{cases} \tag{9}$$

where:

- $\Delta d = \mid d(i, j) - D_{th} \mid$
- θ_{ij} is the orientation of the line segment from robots i to j;
- k is the damping coefficient
- ϵ is a lower bound of Δd. It will be used in order to avoid useless small movements.

In order to overcome the problem in the traditional VFA, we set the attractive coefficient w_a to one tenth of repulsive coefficient k ($w_a = 0.1 \times k$).

When the i-th IoRT robot has more than one neighbor, its new position is calculated as the summation of the position decisions with respect to all the neighbors:

$$\overrightarrow{P_i} = \sum_{j \in N_i} \overrightarrow{P_{ij}} \tag{10}$$

After calculating their new positions, each IoRT robot sends the computed position to the Central Object (CO). Then, CO computes the algebraic connectivity λ_2 of the IoRT robots network according to the formula 3. The central object CO allows each IoRT robot to move to their new positions if and only if $\lambda_2 > 0$. This guarantees that global connectivity is always kept all along the deployment procedure. It is important to note that IoRT robots and CO can communicate to each other through an IoT platform.

The following algorithm summarizes our approach:

Algorithm 1 IoT-based (runs every t units of time)

Phase I: Neighbor Discovery

MyNeighbor = FindNeighbor(RobotId)

Phase II: Compute the position $\overrightarrow{P_{ij}}$ between two robots

Compute $\overrightarrow{P_{ij}}$ using Formula 9

Phase III: Compute the new position $\overrightarrow{P_i}$

Compute $\overrightarrow{P_i}$ using Formula 10

Phase IV: Compute algebraic connectivity

Compute λ_2 of the dynamic Laplacian matrix $L(G)$

Phase V: Deployment

if $\lambda_2 > 0$ **then**
  move to $\overrightarrow{P_i}$
else
  do not move

---

4 ANN-Based Approach

As we mentioned before, the connection to the central object is not always possible. For example, a rescue operation may be difficult after a disaster when the access to the central node is not available. An approach which easily adapts to any type of situation and environment is more than necessary.

To meet this need, we provide an ANN-based technique which can perfectly mimic the behaviors of IoT-based approach. The ANN-based approach is completely distributed and is trained from a set of data. The data set is obtained by using the IoT-based approach and we use backpropagation algorithm to train ANN. The trained ANN is constituted by 2 input units and 1 output unit. The 2 input units are $d(i, j)$ and θ_{ij}, while the output is $\overrightarrow{P_{ij}}$. Therefore, the trained ANN is executed locally for each IoRT robot to control its movement according to its neighbor's distance $d(i, j)$ and angle θ_{ij}. When the position $\overrightarrow{P_{ij}}$ is estimated, the new position $\overrightarrow{P_i}$ of the IoRT robot is computed by using the formula 10. Then, the collective movement of all IoRT robots will allow our trained ANN converge to the desired distance D_{th}. The global connectivity will also keep if our ANN is well trained (i.e. if training error equals zero or near to zero). The algorithm below illustrates our ANN-based approach:

Algorithm 2 ANN approach (runs every t units of time)

Phase I: Neighbor Discovery

MyNeighbor = FindNeighbor(RobotId)

Phase II: Estimate the position $\overrightarrow{P_{ij}}$ between two robots

$\overrightarrow{P_{ij}} = trained\_ann(d(i, j), \theta_{ij})$

Phase III: Compute the new position $\overrightarrow{P_i}$

Compute $\overrightarrow{P_i}$ using Formula 10

Phase IV: Deployment

move to $\overrightarrow{P_i}$

5 Evaluation and Discussion of the Results

In this Section we first describe the simulation parameters and then provide the simulation results of our approaches. We are interested in studying how our approaches converge to the desired distance D_{th} between any pair of IoRT robot (hence to the desired communication quality matching D_{th}). We will see also how the density of

IoRT robots influences the traveled distance of a robot. The importance of taking into account the algebraic connectivity before taking a movement decision will be also highlighted. Our approaches will be compared to the approach described in [21] called hereafter EVFA (Extended Virtual Force-Based Approach). EVFA was designed by its authors to overcome the connectivity maintenance and nodes stacking

Table 1 Simulation parameters

Physical	Propagation model	Two ray ground
	Error model	Real
	Antennas gain	Gt = Gr = 1
	Antennas height	ht = hr = 1 m
	Communication range	250 m
Statistics	Number of samples	100
	Simulation time	3000 s
	Confidence Interval	95%
Mobility	Computation of the new position	see formula (8)
	Damping coefficient k	0.5
	D_{th}	212 m
ANN	Layer number	4
	Input number	2
	Output number	1
	Neuron's number in hidden layers	15
	Desired Error	0.00001
	Max epochs	10000
	Activation function	sigmoid symmetric
	Learning rate	0.2
	Training algorithm	backpropagation
Topology	Topology width	3000 m
	Topology height	3000 m

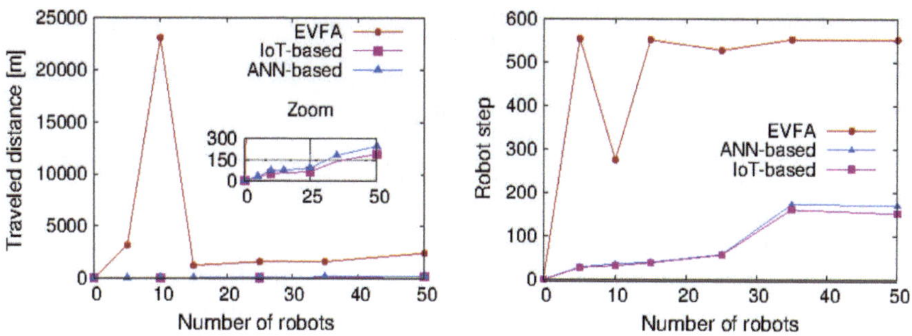

(a) Traveled distance (b) Robot step

Fig. 4 Traveled distance and robot step according to the robots number problems in the traditional Virtual Force Algorithm (VFA). Unlike our approaches, EVFA is based only on the orientation force and the judgment of distance force between node and its one-hop neighbors.

We assess our techniques w.r.t. (i) the algebraic connectivity, (ii) the robot traveled distance, (iii) the average distance, and (iv) the QoS level expressed in terms of RSSI (Received Signal Strength Indicator). Simulations have beeen carried out for a variable number of robots (i.e. from 5 to 50 robots) in an area of 3×3 km.

5.1 Simulation Parameters

All the algorithms in this paper were implemented in version 2.29 of Network Simulator with patch from [14] that reflect a realistic channel propagation and error model. The patch is used in order to provide the effect of interference and different thermal noises to compute the signal to noise plus interference ratio (SINR) and accounting for different bit error rate (BER) to SINR curves for the various codings employed [23]. Table 1 summarizes the parameters used in the simulations.

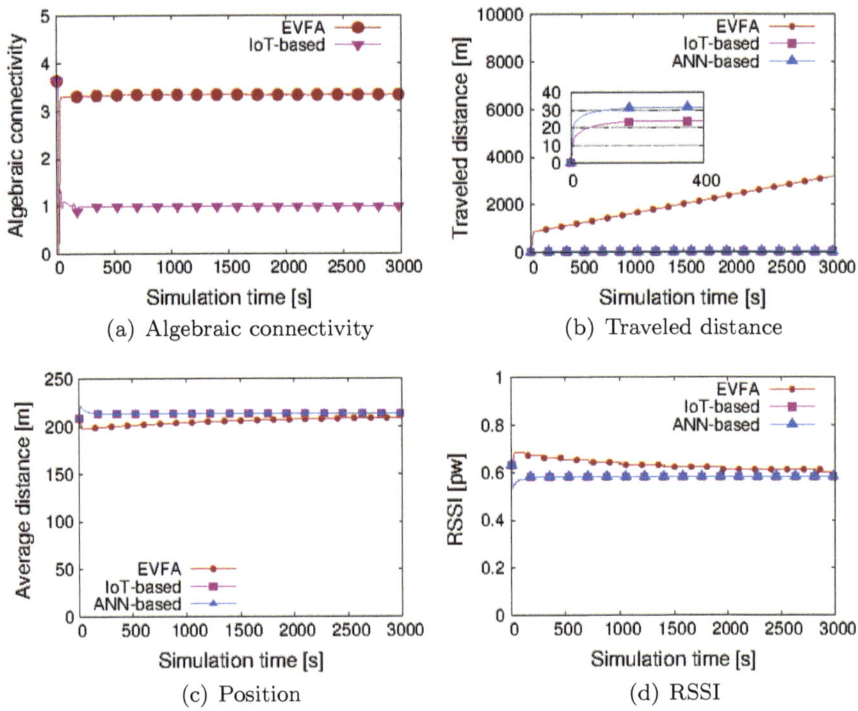

(a) Algebraic connectivity (b) Traveled distance (c) Position (d) RSSI

Fig. 5 Simulation results obtained with 5 robots moving in 3×3 km area

5.2 Simulation Results

All of the following results are the average of 100 times simulations and we assume that the topology is totally connected at the beginning of the simulation.

We can observe in the Figs. 5, 6, 7 and 8(a) that IoT-based approach always kept the global connectivity since it always take account the algebraic connectivity constraint. Unlike IoT-based, EVFA has a connectivity problem when the robots density is low and this can explain why EVFA traveled a lot when the number of robot is less than 15 (see Fig. 4(a)). The goodness of connectivity is observed in EVFA when the robots density is higher (i.e. greater than 15 robots). However, as we mentioned before our goal is not to maximize the algebraic connectivity but just to keep it always greater than zero. This condition is enough to keep the global connectivity.

Figures 5, 6, 7 and 8(c) and (d) illustrate the convergence of our algorithms to the desired distance and the desired communication quality (RSSI) throughout the simulation. We can notice that our approaches converge quickly to the aforementioned parameters which is not always the case for EVFA. We can see also that ANN mimics perfectly the behaviours of the IoT-based approach. This is due to the fact that our neural network has been well trained.

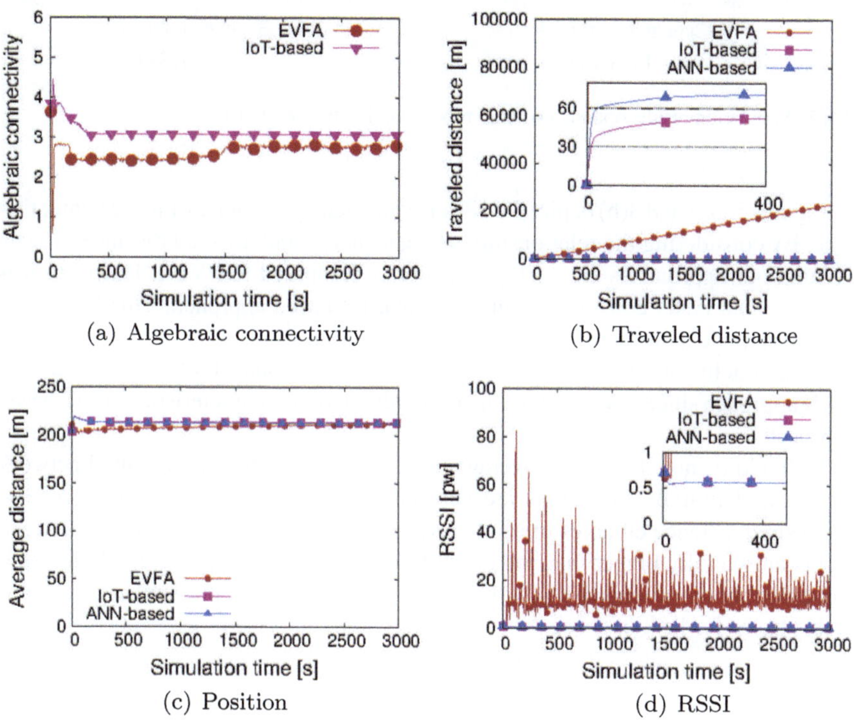

(a) Algebraic connectivity (b) Traveled distance

(c) Position (d) RSSI

Fig. 6 Simulation results obtained with 10 robots moving in 3×3 km area

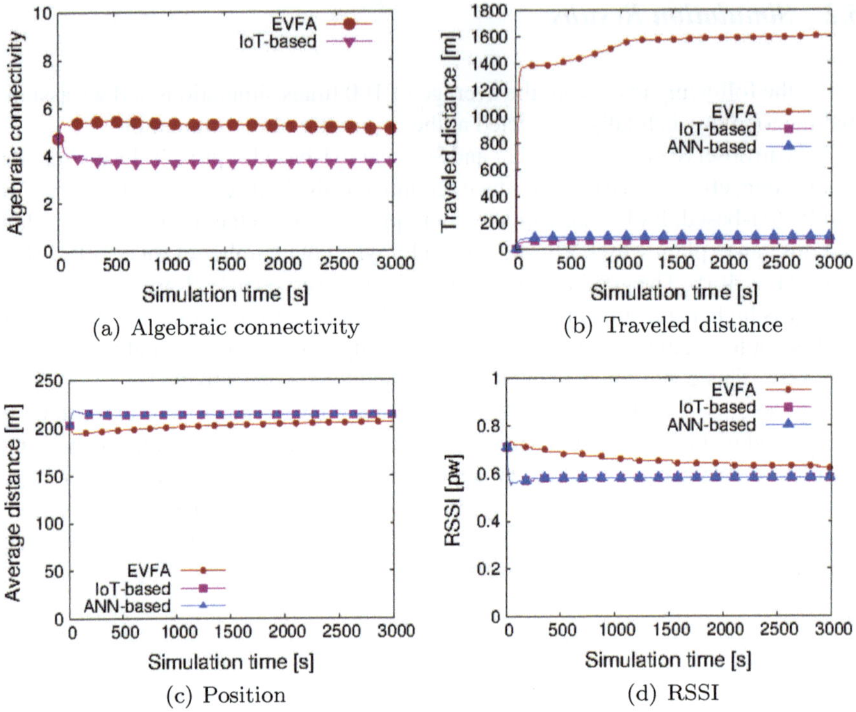

Fig. 7 Simulation results obtained with 25 robots moving in 3×3 km area

Figures 5, 6, 7 and 8(b) depict the distance traveled by a robot during the simulation time. By considering the relationship between energy and traveled distance, we can say that our approaches are energy efficient as compared to EVFA. However, it is observed that ANN consumes a bit more than Iot-based approach. This is because ANN has made a bit more step[1] to converge (see Fig. 4).

As a conclusion, it is worth to say that the global connectivity is reached with ANN approach since it was well trained and inherited the characteristics of IoT-based approach.

It should be noted that the coverage rate strongly depends on the initial network topology. If initially the robots are very close to each other, the convergence to the desired distance enlarges the collective coverage. Else, if the distances between robots are initially bigger than the desired distance, the coverage rate will decrease but the communication quality will improved. This has been proven but we omit it in this paper for lack of space.

[1]Each robot increments its counter when it decides to move. In this paper, robot step is defined as the average of the maximum counter value achieved by one robot during the simulation.

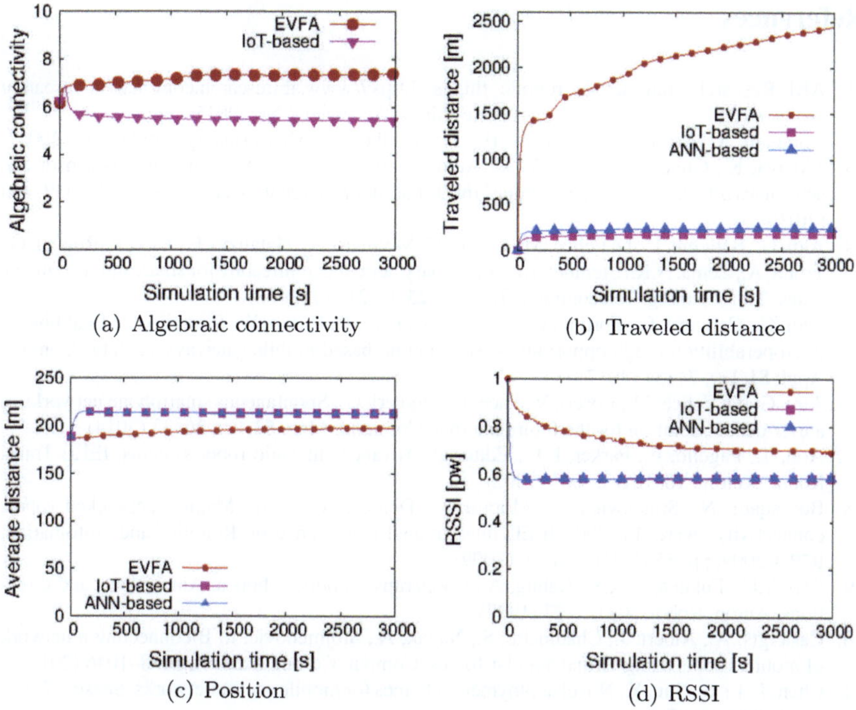

(a) Algebraic connectivity (b) Traveled distance (c) Position (d) RSSI

Fig. 8 Simulation results obtained with 50 robots moving in 3×3 km area

6 Conclusions

In this paper, we implemented a IoT-based and an ANN control scheme to maintain global connectivity among multiple IoRT robots. The proposed approaches tried to capture the trade-off between network coverage and communication quality expressed as RSSI level. The proposed algorithms allow the whole IoRT robot network converges to the desired distance, and hence the desired communication quality. Through extensive simulation we showed that our approaches outperform the EVFA approach proposed in [21], in terms of traveled distance and convergence time. Moreover, our proposed methods always maintain the global connectivity throughout the simulation.

Acknowledgements This work was partially supported by a grant from CPER Nord-Pas-de-Calais/FEDER Campus Intelligence Ambiante.

References

1. ABI Research: Internet of robotic things. https://www.abiresearch.com/market-research/product/1019712-the-internet-of-robotic-things. Accessed 2 Nov 2015
2. Abraham, A.: Artificial neural networks. In: Handbook of Measuring System Design (2005)
3. Akkaya, K., Guneydas, I., Bicak, A.: Autonomous actor positioning in wireless sensor and actor networks using stable-matching. Int. J. Parallel Emergent Distrib. Syst. **25**(6), 439–464 (2010)
4. Aloi, G., Bedogni, L., Di Felice, M., Loscri, V., Molinaro, A., Natalizio, E., Pace, P., Ruggei, G., Trotta, A., Zema, N.R.: Stemnet: an evolutionary network architecture for smart and sustainable cities. Trans. Emerg. Telecommun. Technol. **25**(1), 21–40 (2012)
5. Aloi, G., Caliciuri, G., Fortino, G., Gravina, R., Pace, P., Russo, W., Savaglio, C.: Enabling iot interoperability through opportunistic smartphone-based mobile gateways. J. Netw. Comput. Appl. **81**(12), 74–84 (2017)
6. Aloi, G., Di Felice, M., Loscri, V., Pace, P., Ruggeri, G.: Spontaneous smartphone networks as a user-centric solution for the future internet. Commun. Mag. **52**(12), 26–33 (2014)
7. Arai, T., Pagello, E., Parker, L.E.: Editorial: Advances in multi-robot systems. IEEE Trans. Robot. Autom. **18**(5), 655–661 (2002)
8. Bouraqadi, N., Stinckwich, S., Moraru, V., Doniec, A., et al.: Making networked robots connectivity-aware. In: 2009 IEEE International Conference on Robotics and Automation, ICRA 2009, pp. 3502–3507. IEEE (2009)
9. Cao, Y.U., Fukunaga, A.S., Kahng, A.: Cooperative mobile robotics: Antecedents and directions. Auton. Robots **4**(1), 7–27 (1997)
10. Casteigts, A., Albert, J., Chaumette, S., Nayak, A., Stojmenovic, I.: Biconnecting a network of mobile robots using virtual angular forces. Comput. Commun. **35**(9), 1038–1046 (2012)
11. Chen, J., Li, S., Sun, Y.: Novel deployment schemes for mobile sensor networks. Sensors **7**(11), 2907–2919 (2007)
12. Council: Internet of things council. http://www.theinternetofthings.eu. Accessed 18 Nov 2015
13. Fiedler, M.: Algebraic connectivity of graphs. Czechoslov. Math. J. **23**(2), 298–305 (1973)
14. Fiore, M.: Ns-2.29 wireless update patch. http://perso.citi.insa-lyon.fr/mfiore/research.html. Accessed 19 Nov 2015
15. Fortino, G., Guerrieri, A., Russo, W., Savaglio, C.: Integration of agent-based and cloud computing for the smart objects-oriented IoT. In: 2007 IEEE International Conference on Computer Supported Cooperative Work in Design (CSCWD), pp. 493–498. IEEE (2014)
16. Fortino, G., Trunfio, P.: Internet of Things Based on Smart Objects, Technology, Middleware and Applications. Springer, Berlin (2014)
17. Garetto, M., Gribaudo, M., Chiasserini, C.F., Leonardi, E.: A distributed sensor relocatlon scheme for environmental control. In: 2007 IEEE International Conference on Mobile Adhoc and Sensor Systems, pp. 1–10. IEEE (2007)
18. Gennaro, M.C.D., Jadbabaie, A.: Decentralized control of connectivity for multi-agent systems. In: 2006 45th IEEE Conference on Decision and Control, pp. 3628–3633. Citeseer (2006)
19. Hsieh, M.A., Cowley, A., Kumar, R.V., Taylor, C.J.: Maintaining network connectivity and performance in robot teams (2008)
20. Lee, G., Chong, N.Y.: A geometric approach to deploying robot swarms. Ann. Math. Artif. Intell. **52**(2–4), 257–280 (2008)
21. Li, J., Zhang, B., Cui, L., Chai, S.: An extended virtual force-based approach to distributed self-deployment in mobile sensor networks. Int. J. Distrib. Sens. Netw. **2012** (2012)
22. Parker, L.E.: Multiple mobile robot systems. In: Springer Handbook of Robotics, pp. 921–941. Springer, Berlin (2008)
23. Pavon, J.D.P., Choi, S.: Link adaptation strategy for IEEE 802.11 wlan via received signal strength measurement **2**, 1108–1113 (2003)
24. Petrolo, R., Loscri, V., Mitton, N.: Towards a smart city based on cloud of thoings. In: Proceedings of the 2014 ACM International Workshop on Wireless and Mobile Technologies for Smart Cities, WiMobCity, pp. 61–66. ACM (2014)

25. Qinghai, O., Yan, Z., Xianghen, L., Yiying, Z., Lingkang, Z.: Application of internet of things in smart grid power transmission. In: Twenty-Second Annual Joint Conference of the IEEE Computer and Communications, INFOCOM 2003. IEEE Societies, vol. 2, pp. 1293–1303. IEEE (2003)
26. Riedmiller, M.: Advanced supervised learning in multi-layer perceptrons-from backpropagation to adaptive learning algorithms. Comput. Stand. Interfaces **16**(3), 265–278 (1994)
27. Sabattini, L., Secchi, C., Chopra, N., Gasparri, A.: Distributed control of multirobot systems with global connectivity maintenance. IEEE Trans. Robot. **29**(5), 1326–1332 (2013)
28. Stump, E., Jadbabaie, A., Kumar, V.: Connectivity management in mobile robot teams. In: IEEE International Conference on Robotics and Automation, 2008, ICRA 2008, pp. 1525–1530. IEEE (2008)
29. Wang, G., Cao, G., La Porta, T.F.: Movement-assisted sensor deployment. IEEE Trans. Mob. Comput. **5**(6), 640–652 (2006)
30. Yoon, S., Soysal, O., Demirbas, M., Qiao, C.: Coordinated locomotion and monitoring using autonomous mobile sensor nodes. IEEE Trans. Parallel Distrib. Syst. **22**(10), 1742–1756 (2011)
31. Zou, Y., Chakrabarty, K.: Sensor deployment and target localization based on virtual forces. In: Twenty-Second Annual Joint Conference of the IEEE Computer and Communications, INFOCOM 2003. IEEE Societies, vol. 2, pp. 1293–1303. IEEE (2003)

Transmission Power Control in WSNs: From Deterministic to Cognitive Methods

Michele Chincoli and Antonio Liotta

Abstract Communications in Wireless Sensor Networks (WSNs) are affected by dynamic environments, variable signal fluctuations and interference. Thus, prompt actions are necessary to achieve dependable communications and meet Quality of Service (QoS) requirements. To this end, the deterministic algorithms used in literature and standards, both centralized and distributed ones, are too slow and prone to cascading failures, instability and sub-optimality. Meanwhile, in recent years, cognitive protocols are gradually being introduced. This chapter provides an overview of the Transmission Power Control (TPC) protocols present in literature, categorized as deterministic (proactive and reactive) and cognitive (Swarm Intelligence, Fuzzy Logic and Reinforcement Learning). Only few solutions have considered TPC based on cognitive approaches, including both energy efficiency and QoS management. Our review identifies key shortcomings in deterministic TPC, pinpointing the benefit of the emerging methods based on computational intelligence.

1 Introduction

Transmission Power Control (TPC) has been used already for many years, initially in cellular networks for the purpose of energy and spectrum efficiency. Base Stations (BSs) control and manage the transmission power for their subscriber units by instructing them to raise or lower the level in 1 dB steps, using the forward control subchannel [44].

Many years later, TPC has been introduced in Wireless Sensor Networks (WSNs) whereby energy and spectrum efficiency are particularly critical. Energy consumption has been from the beginning a crucial problem in WSNs, considering that sensor nodes are small devices with limited resources and energy supply. The battery

M. Chincoli (✉) · A. Liotta
Electrical Engineering Department, Eindhoven University of Technology,
PO Box 513, 5600 MB Eindhoven, The Netherlands
e-mail: m.chincoli@tue.nl

A. Liotta
e-mail: a.liotta@tue.nl

R. Gravina et al. (eds.), *Integration, Interconnection, and Interoperability of IoT Systems*, Internet of Things, DOI 10.1007/978-3-319-61300-0_3

replacement or recharge is not always practical and is even impossible in harsh environments [29]. This is the reason why the energy needs to be spent carefully during a sensor node operational time, which is often expected to be in the order of years. The energy consumption of sensor nodes has been tested in different studies, analysing the impact of sleep techniques and protocols on the WSNs. One example is the work in [30], where the energy variation is investigated under the use of a TPC protocol in multi-hop networks. Moreover, since WSNs are meant for providing services to users in the context of Internet of Things (IoT), energy efficiency is again important to maintain connectivity in the network [14, 15, 47].

Concurrently, WSNs have to also address Quality of Service (QoS), in relation to network reliability, by satisfying performance constraints of specific applications. To this end, distributed and reconfigurable scheduling acquire considerable importance, as well as intelligent techniques for anomaly detection [2–5, 13, 16]. At the same time, TPC protocols are optimized cooperating with other protocols and techniques (e.g., routing, localization) [23, 24, 27, 28].

The importance of energy in WSNs is exemplified in the work undertaken collectively in S-MAC [55], Timeout-MAC (T-MAC) [51], Pattern-MAC (P-MAC) [57], Reinforcement Learning-MAC (RL-MAC) [38] and Q-Learning-MAC (QL-MAC) [17, 18]. All these protocols focus on energy efficiency but pay lower attention to QoS management. They employ a frame-based structure, which means that the time is organized in frames that are further divided in smaller portions of time, called slots. Each slot can be used for sleep state, where the transceiver is turned off, and active state, where both transmission and reception can be scheduled. Among the aforementioned protocols, a novelty is introduced in both RL-MAC and QL-MAC by using a cognitive approach, namely Q-learning, a simple and powerful Reinforcement Learning (RL) algorithm [50]. The former has been able to improve the throughput and energy consumption compared to S-MAC and T-MAC, by rewarding the system with the node's internal information and the indirect information of the other nodes status. Similarly, QL-MAC, following the above approach, obtains better performance in terms of Packet Delivery Rate (PDR) and average energy consumption, in comparison to S-MAC, T-MAC and CSMA/CA.

The specific objective of this chapter is to review the literature on the most notable TPC protocols in WSNs, differentiating them between deterministic and cognitive solutions. The former are based on various observations and thresholds which, by default, are not capable of taking in consideration the dynamic nature of the WSN environment. What is worse, the competitive and egoistic nature of those protocols is not compatible with the requirements of high-density wireless communications, whereby the increased power leads to a cascading effect across the network, leading to energy and spectrum wastage.

For this reason, another interesting research direction in WSNs is gaining attention by the use of Machine Learning methods (e.g., Reinforcement Learning, Swarm Intelligence) to optimize various aspects of WSNs [19, 34, 35]. Such methods fall under the category of cognitive protocols. These are recognized to be very effective at predicting network metrics under dynamic conditions [36]. Some methods are even suited to distributed architectures [20]. On the other hand, not all cognitive solutions are viable in WSNs, especially those involving complex computation or data caching.

The remaining of this chapter is organized as follows. Section 2 examines the taxonomy of TPC protocols, dividing them in the two broad categories of deterministic and cognitive. Related works are discussed in Sect. 3, specifically how other authors have selected and classified studies in literature. Sections 4 and 5 present the description of some examples of proactive and reactive protocols respectively. Next Sects. 6, 7 and 8 introduce the three prominent cognitive protocols, Reinforcement learning, Fuzzy Logic and Swarm Intelligence. Lastly, Sect. 9 concludes the chapter with final observations and remarks.

2 Taxonomy

TPC algorithms have been used already for decades and many contributions can be found in the state-of-the-art. Among them, we can find also surveys that categorize the area according to a variety of criteria [12, 43]. Protocols typically focus on one or more OSI layers, are either centralized or distributed, and either use or do not information from the network. Most of them target energy efficiency, ignoring QoS guarantee that is also very important in WSNs as explained in [7].

Initially, most studies in the literature proposed protocols based on deterministic algorithms, with few having a predictive nature. The general idea is to monitor the link quality between nodes and use it as metric to reduce energy consumption. The main difference among various approaches lays on the method used to address efficiency. There are diverse metrics for link quality evaluation; different means to perform data analysis; distinct ways to layer the architecture; dissimilar energy targets (i.e., local or global energy consumption in a network); and a variety of sensor sleep techniques (more on this in Sect. 3).

In this chapter, we propose a new methodology to distinguish TPC protocols based on deterministic and cognitive algorithms (Fig. 1). We group deterministic protocols in proactive and reactive approaches and categorize cognitive solutions as three unsupervised machine learning categories, namely Reinforcement learning, Fuzzy Logic and Swarm Intelligence. Each protocol of the aforementioned classification can have common goals, aiming either for energy efficiency, QoS guarantee or both. In Table 1, we summarize all the protocols discussed in this chapter, specifying the category and class in which they belong, and the target appointed.

The terms *proactive* and *reactive* are more common in the context of routing protocols [1]. We borrow this nomenclature to better categorize TPC protocols. Thus, proactive algorithms calculate and estimate the transmission power to be used in subsequent transmissions. This can be done by means of mathematical modelling or via the exchange of beacon probes during an initialization phase. In this way, some protocols can be predictive as well as adaptable based on the subsequent transmissions, where a node receives feedback subject to internal information (e.g., statistical

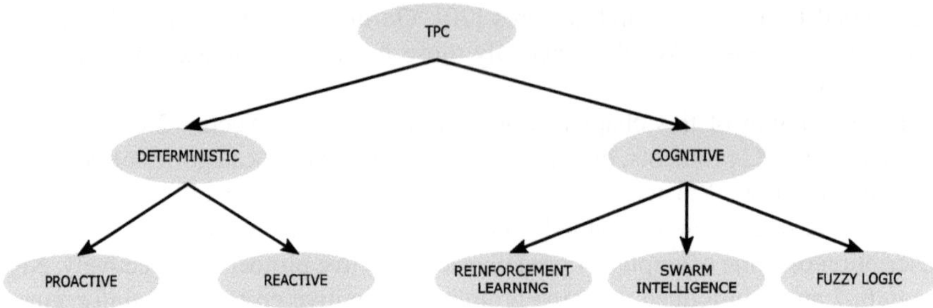

Fig. 1 Taxonomy of transmission power control in wireless sensor networks

Table 1 Protocol classification

Protocol	Category	Class	Target
PCBL [49]	Deterministic	Proactive	Energy efficiency
ATPC [37]	Deterministic	Proactive	Energy efficiency
AODTPC [40]	Deterministic	Proactive	Energy efficiency
ART [22]	Deterministic	Reactive	Energy efficiency and QoS
ODTPC [25]	Deterministic	Reactive	Energy efficiency
PSO [52]	Cognitive	Swarm intelligence	Fusion error probability
Collotta et al. [10]	Cognitive	Fuzzy logic	Energy efficiency
Martinez et al. [39]	Cognitive	Fuzzy logic	Energy efficiency and connectivity
Pandana et al. [42]	Cognitive	Reinforcement learning	Energy efficiency and QoS
Kobbane et al. [26]	Cognitive	Reinforcement learning	QoS
Lin et al. [33]	Cognitive	Reinforcement learning	QoS
Gummeson et al. [21]	Cognitive	Reinforcement learning	Energy efficiency

computation) or/and other nodes information. Thereby, a node is ready to use a predetermined transmit power level for each neighbour. On the other hand, reactive TPC protocols do not try to anticipate power levels ahead of each transmission. Whilst this saves energy by not injecting overhead in the network, yet since the nodes ignore the network, power levels are set based on the most recent local information, which often leads to incorrect decisions. Indeed, reactive protocols use algorithms to calculate the transmission power on demand, when a packet is going to be transmitted or after a reception. Furthermore, the initial power level is unconditionally set to the maximum or to a default one.

With regard to cognitive protocols, these rely on Computational Intelligence (CI), which is particularly indicated for systems that are complex and dynamic such as WSNs. CI brings flexibility, adaptability and robustness to address issues such as topology variation, communication failures and scenario changes [31]. Key CI paradigms include Neural Networks, Fuzzy Logic, Evolutionary Algorithms, Swarm Intelligence, Artificial Immune Systems and Reinforcement Learning. CI paradigms have been used to improve the reliability of WSNs considering the following challenges: Design and Deployment, Localization, Security, Routing and Clustering, Scheduling and MAC, Data Aggregation and Fusion, and QoS Management. Most studies have focused on the routing and clustering aspects of WSNs, while very little is known about the QoS Management issue. Routing and clustering have been studied using Reinforcement Learning and Swarm Learning algorithms, followed by Evolutionary algorithms. Instead, while Fuzzy Logic has been the preferred choice to address QoS management, a most recent and innovative solution has been proposed by Zhao et al. in [56] for dense wireless networks. They apply Transfer Learning for a coordinated and distributed approach targeting QoS performance. Two models of cooperation management are presented, combining distributed Reinforcement Learning and Transfer Learning.

Among the cited paradigms, TPC protocols are inherent to the QoS Management, and Scheduling and MAC challenges.

3 Related Works

Thus far, a number of studies have reviewed TPC in WSNs, so it is worth going through the different viewpoints, before introducing our new perspective. The classification by Pantazis et al. (2007) [43] focused on energy reduction and lifetime improvement in sensor nodes. Furthermore, the authors underline the need to fill a gap in literature, including QoS guarantee in TPC protocols, which has significant implications to the reliability of WSN applications. As far as the power management classification is concerned, existing mechanisms are divided as passive and active ones. The former regulate the hardware components of the sensor nodes (e.g., processor, memory, transceiver, A/D converter), assigning periods in which they are turned on or off. Passive techniques can be further distinguished into physical layer power conservation, fine-grain power conservation and coarse-grain power conservation, depending on the layer where the period assignment management is done. Specifically, each technique involves the physical, MAC and upper layers respectively. On the other hand, active techniques are classified in MAC, Network and Transport layer, according to the layer in which they operate.

Passive physical protocols aim at improving the energy efficiency by calibrating the duration and frequency of the active state. In other words, the system should enable active states only when is strictly necessary, avoiding idle cycles. Different methods have been pursued to achieve this goal, such as Dynamic Voltage Scheduling (DVS) and Dynamic Power Management (DPM). DVS adjusts the speed of the processor to the minimum required clock frequency, to improve the CPU energy efficiency. Acting on the voltage speed processors, their speed is modified by varying the operating frequency along with the power supply. The main problem comes with the difficulty to foresee the processor activity in real-time.

By contrast, DPM focuses on switching the sensor node components on and off when necessary. Although beneficial in some cases, exploiting sleep states becomes disadvantageous in others. Because moving from an active to a sleep state and *vice-versa* consumes resources and power, it is convenient to switch to energy saving mode only if the time spent in the sleep state is higher than a threshold. More energy-efficient solutions are offered: parallel computation partition among neighbour nodes by lowering their voltage supply levels and the clock frequency; adaptable transmission power per link by adjusting either the output transmit power or adding the Forward Error Correction (FEC) module; minimization of the operations in Digital Signal Processor (DSP) systems by exploiting time-varying signal processing.

With respect to Fine Grain Protocols, the Request-To-Send/Clear-To-Send (RTS/CTS) handshake for CSMA/CA is commonly selected. For example, in [48], two separate channels are used, one for Data and the other for control. Receiving nodes transmit a busy tone in the control channel to signal the presence of a communication. The consequence is that both nodes that have to transmit and those without pending data can turn to sleep mode.

Protocols that belong to Coarse-Grain group are categorized in Distributed (CG-DA) and Backbone-based (CG-BBA) approach. These differ on the topology based on the architecture where the communications rely on. In CG-DA, the nodes are distributed, while in CG-BBA are disposed in a backbone. S-MAC is a protocol under CG-DA, where the sensor nodes have a synchronized sleep period and are allowed to transmit during the wake-up mode [55]. In this way, collisions and overhearing are avoided, in addition to the energy savings during periodic periods of sleep mode, improving the energy consumption. Another interesting protocol is Geographical Adaptive Fidelity (GAF) in [53], in which the sensor nodes are aware of their positions through the help of the Global Positioning System (GPS). The concept of such protocol is to simplify the routing paths assuming that a subset of the nodes is enough to keep the network fully connected in specific grid areas.

A similar protocol, belonging to CG-BBA, has been proposed in [6]. The main difference with GAF is that the nodes do not know their geographical position but broadcast messages are used to react to topology changes. The idea behind the protocol is to define the number of active nodes to join a backbone that is considered sufficient for network traffic forwarding.

Turning now to active protocols, many are part of the MAC layer and are based on Multiple-Access techniques in order to improve the transmission scheduling and avoid the overhearing. For instance, three-layer handshake RTS/CTS/DATA or its modified version with ACK, RTS/CTS/DS/DATA/ACK, can be used. Moreover RTS/CTS can be accompanied with a busy tone, or even a dual busy tone as in Dual Busy Tone Multiple Access (DBTMA), so as to control the transmission power of each transmitter.

Schemes at routing layer are divided in Power-Aware and Maximum Lifetime Routing (MLR). In the former, the routing path is generated minimizing the energy costs with the help of five metrics: minimization of the consumed energy per packet, maximization of time to network partition, minimization of variance in sensor node power levels, minimization of the cost per packet and minimization of the maximum sensor node cost. Among the listed protocols, Power Management for Throughput Enhancement in Wireless Ad-Hoc Networks is particularly interesting. It is based on the effect of the transmission power reduction on the sensor nodes. Whilst the connectivity range is shrunk, with beneficial consequence on the interference, the traffic is relayed over multiple hops before reaching the destination, producing a higher number of transmissions in the network and a counterproductive interference. For this reason, the bounder between the two cases is analyzed.

As regards MLR protocols, the main target is the generation of a topology in such a way the network lifetime is maximized. For example, routing in [11] uses a virtual backbone for computing and updating routes (i.e., shortest paths). Finally, few Transport layer protocols are discussed, such as Pump Slowly, Fetch Quickly (PSFQ) and Event-to-Sink Reliable Transport (ESRT).

Another work that examines TPC in WSNs is [12], focusing on adaptive communication techniques in IoT systems. The authors invite the reader to expand such techniques in every OSI layer and component of WSNs. They illustrate a conceptual scheme for adaptive IoT communications to deliver a threefold message. Firstly, WSNs are integrated with the Internet, their components are heterogeneous and must be able to communicate with external agents, such as WiFi and cellular devices. Secondly, sensor nodes can be either end nodes, routers or cluster heads. Simplifying the manufacture, nodes have the same hardware but, depending on their role, only some features are enabled. Thirdly, the component stack shows that all the layers, excluding the application one, are interconnected with each other, exchanging relevant information and obtaining as result an adaptive system.

The physical and MAC layers are crucial for spectrum sensing, decision and reconfiguration (e.g., the transmission power and channel). Four main spectrum sensing techniques are discussed and compared: matched filter, cyclostationary feature, energy detection and eigenvalue-based detection. Such capabilities are provided by the physical layer and controlled by the MAC layer. For this aspect, the sensing mode (proactive and reactive) and duration must be taken into account with consequent pros and cons. The advantage comes from getting more data from other devices, despite the local information that a node is able to collect. Instead, sensing the spectrum improperly has drawbacks on network performance. Thus, cooperative sensing has an important role and its architecture, methodology and scheduling need to be deployed carefully depending on the network condition and requirement. The architecture can be either centralized, if a central node collects all the data, or clustered, where the network is divided in clusters managed by a cluster-head. The methodology is relevant especially for the latter case, for instance to decide certain actions within a cluster. Lastly, the sensing scheduling can be either random or follow Time Division Multiple Access (TDMA). The choice and combination of different aspects just described bring to different results.

4 Proactive TPC

In this section, we provide the description of some important examples within the category of proactive TPC protocols.

Transmission Power Control with Blacklisting scheme (PCBL) is proposed in [49], differentiated in two versions: optimization prior to routing and on-demand optimization for long-lived routing. Initially, in the former approach, a collection of link statistics is done, in terms of Packet Reception Ratio (PRR), for every transmission power level. The best transmission power is chosen per link, as the minimum that respects a link quality requirement, where else the links under a blacklist threshold are considered unreliable and so blacklisted. Among the selected power levels for unicast communications, the maximum one is used for broadcast communications.

Alternatively, the second approach is offered to reduce the overhead in the network. Indeed, only the maximum transmission power is used in the initialization phase, the reliable links are identified and the remaining ones are blacklisted as unreliable. This information is used by the routing protocol to find a delivery path and unicast transmission power are set.

PCBL provides energy-efficient link quality control with minimal channel interference, as well as new network topologies with more accurate and robust links.

Another proactive protocol can be found in [37], where an Adaptive Transmission Power Control (ATPC) is presented and analyzed. Following thoroughly link quality analysis in point-to-point communications between nodes in a linear topology, the authors discovered a correlation between the RSSI and transmission power that can be approximated as linear. Pursuing such behaviour, a linear regression with least square error is adopted, for prediction of the transmission power, feeding the algorithm with RSSI values. All the transmission power levels are stored in a neighbour list and, indeed, assigned to each neighbour. At the beginning, the neighbour list is filled by choosing the power, associated to a RSSI value that satisfies a given link quality threshold. RSSI values per power are collected in the initial phase where nodes send broadcast messages at every transmission power; thus the receivers memorize the RSSI associate to individual senders. As a result, the coefficient that dictates the translation of the curve in the linear regression is adapted in time using the latest transmission power and RSSI pairs from receivers' feedback. Adjustments in the transmission power are necessary when relevant changes in the RSSI are detected comparing real-time values with two thresholds. The paper reveals that the protocol performs well in scenarios where individual communications are isolated per time slots, as for instance in Time Division Multiple Access (TDMA).

Predictive RSSI analysis is done in [40] using the Kalman filter for the Adaptive On-Demand Transmission Power Control (AODTPC) protocol. The idea here is to use the proper transmission power level, based on the wireless channel condition, before the transmission of a packet. Differently than [22, 25, 37], AODTPC does not rely exclusively on feedback from the receiver, but predicts RSSI values beforehand, feeding the Kalman filter with the received power via the theoretical log-normal shadowing model and ACKs from the receiver for error correction. AODTPC can also estimate future values without ACKs. After predicting RSSI values, the protocol chooses the transmission power level following the rules of ODTPC in [25]. The results show that the trend of RSSI in AODTPC is indeed slightly different than ATPC and ODTPC because, as the authors say, the model catches better the channel variability, even though sometimes the estimation error increases. However, the energy consumption in AODTPC is lower than the other two protocols for two different scenarios, where the RSSI low threshold is changed.

5 Reactive TPC

Alternatively to proactive TPC, we collect other solutions in this section belonging to reactive protocols.

Adaptive and Robust Topology control (ART) [22] aims to reduce network contention and energy consumption. The protocol calculates, within a window of N slots, the packet failures as Link Quality (LQ) at the transmitter side, based on the Acknowledgment (ACK) losses over the successfully transmitted packets. The failures are compared with two thresholds, low and high. If the metric is lower than the low threshold thus the system is achieving better performance than expected and enters in a trial state. Here, the transmission power is decreased and the packet reception is analyzed in real time. In case the number of failures equals the low threshold, while the window is possibly not completely filled yet, the old transmission power is restored; otherwise the new transmission power is confirmed and the trial state is left for the steady state. Differently, if the packet failure is above the higher threshold, ART uses a flag that monitors the contention in the network. This means that the transmission power is increased in the situation where the contention is low, assuming that the weak received power is attributed to the wireless channel. Instead, the transmission power is decreased in the opposite situation, where the network is suffering from high contention, hence it is necessary to diminish the interference. Alternatively, if the number of failures is between the two thresholds, the system is stable in a steady state, and the transmission power remains constant. The results confirm that the power consumption is reduced in comparison to the scenario where only the maximum transmission power is used. Moreover, the PRR in ART is analyzed and compared with the scenarios using Max-Power and PCBL protocol [49]. PRR, as the ratio of the ACK received and the successfully transmitted packets, is shown to be higher than Max-Power and similar to PCBL.

ART has also been studied and compared to scenarios having homogeneous transmission power and variable interference [8, 9]. The authors have compared the performance of a network, using ART and a homogeneous, constant transmission power, in terms of PRR and latency. In case of homogeneous transmission power, different scenarios have been simulated, changing the level of interference by modifying the density of the network, as number of nodes per area, and varying the distance of the interferer nodes from the probe nodes. From the results, an intermediate transmission power level reveals to provide better performance than the case in which the power is set close to the maximum or minimum levels.

A similar mechanism of threshold comparison for the transmission power updates is done in [25]. In this work, the protocol On-Demand Transmission Power Control (ODTPC) is evaluated. ODTPC calculates the transmission power when a packet is ready to be transmitted. It is based on two phases: large-scale transmission power control (L-TPC) and small-scale transmission power control (S-TPC). L-TPC is enabled when a transmitter is ready to send a packet to a neighbour for the first time. In such case, the highest transmission power is used and the receiver approximates the proper transmission power level, plus a margin, exploiting the RSSI. The resulting value is sent to the transmitter that approximates its appropriate transmission power level. After this operation, the transmitter switches to the second phase S-TPC related to the same neighbour. When the neighbour has received a packet, it replies with an ACK, including its RSSI. Then, the transmitter compares the RSSI with two thresholds. If the RSSI is below the low threshold, the power level is increased by a fixed step. Otherwise, if the RSSI is above the high threshold, the transmission power is decreased by the same fixed step. ODTPC has been experimentally compared to ATPC and PCBL. The authors prove that their protocol achieves better PRR than PCBL and ATPC for both single and multiple data flow scenarios. In the first scenario, the transmission power level chosen is minimum in the case of PCBL, since a PRR above 95% is considered acceptable. At the same time, the power used in ODTPC is lower than in ATPC. In the second scenario, ODTPC outperforms PCBL and ATPC in terms of PRR and energy consumption.

6 TPC Based on Swarm Intelligence

Swarm Intelligence is an approach aimed at reproducing the behaviour of animal swarms in nature. For example, bees, ants, birds, just to name a few, have the ability to be coordinated in movements and tasks. This phenomenon is fascinating and powerful and, for this reason, a lot of effort goes in the same direction to replicate swarm capabilities in computational intelligence [41]. Different studies have been carried out also in the context of WSNs [32, 45] but only few of them have been considered for TPC.

In [52], Particle Swarm Optimization (PSO) is used for optimal power scheduling in a network having as main target the transfer of information from sensor nodes to a fusion center. In other words, the optimal power scheduling improves the mean squared error performance compared to the case where only a homogeneous transmission power is used. The results are shown to be better when a large number of sensor nodes compose the network or when the majority of the link quality is good.

The authors have thoroughly studied the optimal policy when the observations are independent and identically distributed (i.i.d.), correlated by a small coefficient, and correlated at different grades. The power allocation problem for the first two cases is formulated analytically, except for the fact that when the correlation is small, an upper bound is used for the fusion error probability. When local observations are i.i.d., the solution to the optimal power allocation problem is found when the total power spent by the entire WSN is minimized and the fusion error probability is under a certain threshold. Alternatively, in the second case, the fusion error probability bound is targeted to be minimized.

Different conditions apply when the observations are arbitrarily correlated, as it is not possible to find a closed form for the problem, thus it is solved numerically by a stochastic evolutionary computation technique based on PSO. This is possible when the PSO is used for an unconstrained optimization problem. Otherwise, as in the case of the optimal power allocation, which is a constrained optimization problem, the exterior penalty function can be used to transform it into an unconstrained optimization problem.

PSO terminology is used to describe the elements of a WSN. A sensor node is considered as an agent or particle and is identified with its own location. The entire collection of agents is represented as a swarm. In the initialization phase, it is important to define the parameters to be optimized, the range of the space where the solution can be found, and the fitness function. The latter provides in output a single number representing the goodness of a given solution and needs to be related to the importance of each aspect being optimized. During the iteration of the algorithm, each agent has in return the best location and the maximum allowed velocity to move toward it. In parallel, the algorithm provides with the best location that the entire swarm is indicated to move. Both locations are given as output of the fitness function. The individual best location is iterated by different steps, choosing the one corresponding to the minimum fitness value. The velocity is determined according to the distance between the current location and the next best individual location and global location. At this point, the agent is moved to its new location. Afterwards, the values of each parameter are repeatedly updated until the termination criteria are met.

At the end of this process, an optimal solution to the exterior penalty function can be obtained. When the observation noise is i.i.d., the function is strictly convex and a global minimum exists which certainly converges to the optimal solution of the original problem, considering a series of assumptions.

The results show that the nodes that have bad observations and/or link quality are turned off for better performance and energy consumption. The energy waste is even improved when many nodes are part of the network as well as many links have good quality, and is reduced with respect of uniform power allocation. In such manner, it is not necessary to collect observations from a large quantity of nodes and so a big part of them can be switched into sleep mode. Better performance have been found also using the PSO method compared to the situation with independent observations.

7 TPC Based on Fuzzy Logic

Fuzzy Logic (FL) has been used widely for WSNs, because of the simplicity of the algorithms that diminish the complexity of the computation in constraint devices. Differently than the binary approach in the classical logic where a proposition is either true or false, in Fuzzy Logic there is fine granularity between 0 (false conclusion) and 1 (true conclusion). The philosophy in FL is that every existent element are part of either a set or its complement. The member of a set X is been assigned a degree of membership to a subset Y. Algorithms are expressed by IF-THEN statements and operators known also in the boolean terminology (i.e., AND, OR, NOT).

Among the studies in literature, [10] applies FL to WSNs for energy efficiency. The topology of the network consists of Network Controllers and Field Devices. The Network Controller has the role to regulate the power consumption of the Field Devices, which is done with the help of two parallel Fuzzy Logic Controllers (FLCs). Whereas the former takes in input the battery level and the throughput to workload ratio, giving in output the sleeping time, the latter is responsible to assign the power level taking in input the battery level and the link quality. The throughput is the sum of both periodic and aperiodic messages sent by a device, instead the workload is the total number of packets that the device has to send. The link quality is provided by the hardware of the device, whose value is between 0 and 255. Both FLCs use three membership functions (i.e., Low, Medium and High) for the inputs and outputs. Thus, the inference rules are done by the combination of the input membership functions and the assigned one to the output parameter. Moreover, the degree of membership for the input parameters is represented by a triangular-shape function. The output value is calculated by defuzzification using the centroid mechanism. The results show that using the parallel fuzzy scheme compared with a fixed sleeping time and transmission power, and the approach proposed in [46], the battery has a longer lifetime.

Martinez et al. in [39] have addressed the problem of connectivity in WSNs together with the energy consumption. The idea is to have good connectivity by choosing a k-value at each node, which is used to obtain a necessary number of neighbours for a k-connected network. The k-value is the range of the targeted node degree within a tolerance interval that is adjusted dynamically modifying the transmission power level of the node. In addition, the battery level is also considered in the system in order to be kept around a safe value. This is done via a self-adaptive strategy, based on fuzzy logic, with the support of a control system. The latter is managed in two feedback loops and divided by three blocks: monitoring, reasoner and actuator. The monitoring block analyses the current state of the node and calculates the error variables; the reasoner takes the decision once triggered by the previous block; and the actuator implements the desired changes in the node. Each module has different tasks that are triggered following some fuzzy rules. In combination, the first feedback manages the transmission power of a node based on the real and targeted number of its neighbours. The second one is responsible for the energy consumption, considering the battery level as well as the targeted quantity of peers that can be reached by a node. For instance, if the energy is low, then the communication range is decreased by means of reducing the transmission power, lowering the number of connected nodes. It is important to mention that the neighbours for each node are discovered by a parallel protocol to a standard routing protocol (e.g., Ad hoc On-demand Distance Vector, Link Quality Routing Protocol- AODV, LQRP).

The aforementioned system has been tested through experiments in an outdoor environment. The experiments are distinguished by choosing different values for each parameter. The authors have shown that for certain experiments the results are very promising, obtaining an optimal Packet Delivery Ratio (PDR) (e.g., above 96%) and reaching it steadily in few iterations. Furthermore, when the network reaches the steady state, the nodes have different node degree, and when there is no change in the transmission power, the node degree also changes. This assumes that the wireless channel is dynamic besides the transmission power chosen by the nodes of the network.

In [39], there is also a discussion regarding the strategy in choosing the transmission power that depends on the target of the system. If at least the 87.8% of the target preference falls into the energy saving than the connectivity, the fixed medium transmission power should be preferred. Otherwise, if the energy efficiency is the least relevant choice, up to 45.5%, the fixed transmission power should be set to the maximum. In all the other cases, the transmission power should be self-adaptive.

8 TPC Based on Reinforcement Learning

Reinforcement Learning (RL) is an appropriate solution for scenarios where the environment is dynamic and unknown, and where the data are collected on-line. The system learns how to reach a certain target by exploring and exploiting actions (chosen based on a policy [50]), in a given state that represents the status of the environment, and receiving rewards to understand whether the choice made has been positive or negative. The goal of RL is to maximize the reward in the longer term, considering then value functions (i.e., state-value and action-value functions).

RL has been applied for many problems such as in Medium Access Control (MAC), routing, rate control, cooperative communications, sensing coverage and task scheduling [54].

The authors of [42] have used an algorithm to maximize the average throughput per total consumed energy in a point-to-point communication and multi-node network. The authors obtain a near-optimal control policy, in comparison to the optimal policy based on solution of the Bellman optimality equation. The results from both policies are furthermore compared with a simple policy, where each transmitter uses the highest modulation possible, while maintaining the SINR of a link for the chosen modulation. The three elements of the algorithm, states, actions and the reward, are associated as follows: the first one is a combination of the channel gain and the number of packets in a queue; the second one is a combination of the transmit power level and modulation level that have been used for a transmission; and the latter is the average throughput per total consumed energy. In both the point-to-point and multi-node scenarios, the learned throughput outperforms the simple policy. In addition, in the former case, the plots depict that, when the buffer is full or the channel gain is low, the transmission power is high. In parallel, fixing a value for the channel gain, as the buffer content increases, also the modulation level rises. As last note, the authors underline that another important aspect of RL algorithms is the transition time before reaching a stable level in the performance. Indeed, this is strictly related to the number of states and actions, thus using fewer of them implies faster transient times.

In [26], a dynamic centralized power control mechanism for Multimedia WSNs (MWSN) is proposed. Through a Markov Decision Process (MDP), each sensor node has a quantized number of battery levels represented in states. At each state, a selection of transmission power levels can be chosen. The transition between states occurs based on the next state's conditional probability knowing the previous state and the transmission power.

The target of the method is the maximization of the throughput, which is calculated through an increasing function of the Signal Interference Noise Ratio (SINR), specifically adopting the Shannon function. Each sensor gains a reward, equal to the throughput, after choosing a certain transmission power level, in order to receive an evaluation of the taken action's effect over the battery lifetime. The reward is taken into account in defining the optimal policy that provides the transmission power to use, which maximizes the Bellman optimality equation.

The method is compared to other policies: Not transmitting (NTRANS), transmitting at the highest available power (HighP), and transmitting at the lowest available power for the first half of the maximal battery lifetime and at the highest available power for the rest (HybridP). The optimal policy outperforms HybridB and HighP, in terms of expected battery lifetime, while NTRANS increases linearly as the battery lifetime is extended. Moreover, the optimal transmit power for one sensor is proportional to the battery state when the time is approaching the maximal lifetime. Also when the optimal policy considers the decision that many sensors should take, the optimal strategy tends to use higher transmit powers in higher states.

In [33], instead, an analysis of different algorithms is done aimed at electing the most suitable for multi-hop wireless networks. The study poses the necessity of finding solutions for delay-sensitive applications in such a way that traffic is reliably delivered, therefore the adoption of a distributed protocol is preferable. Such target is approached via an autonomic and distributed joint routing and power control mechanism that is formulated as an MDP.

The contribution of this work includes the procedure to define the optimal policy for each node of the network by assuming that the network dynamics are known. Then, since the wireless channel is very dynamic and unpredictable, and thus the global state transitions probabilities in MDP are unknown, an online learning algorithm is used together with a mechanism to reduce information exchange in a distributed architecture. In the MDP formulation, the states, actions, state transitions and rewards are defined. The states are the expression of the wireless channel dynamics and the packet queue occupancy; the actions consider the routing path and the transmission power level selection; the state transitions satisfy the Markovian property; the rewards represent the expected number of received packets within a delay constraint. The actions taken in a certain state are chosen based on the optimal policy (i.e., one that maximizes the expected discounted reward).

The problem treated in this work considers the behaviour of all the nodes in a network, hence the number of states and actions are large, making the complexity of the system impracticable for a centralized solution. Moreover, a factorization of the state transition probability is done; thus one transition at the nodes in the h-th hop is dependent only from the state and action of the nodes in the *(h-1)*-th hop. Subsequently the distributed state-value function is defined as the expected accumulated discounted reward and is computed solving the dynamic programming (DP) problem, locally, in a hop-by-hop manner.

At this point, the technique is enriched with the addition of online learning. The value-function is modified following the temporal difference between its current and previous estimation, and the policy update is implemented using the actor-critic (AC) method. The critic is associated to the value-function that is used to strengthen or weaken the tendency of choosing a certain action, while the actor is the policy structure, a function of state-action pair, which indicated the tendency of choosing action a at a state s.

The last step focuses on the trade-off between the performance and the information exchange overhead. Better results can be obtained with more data exchange to elaborate more precise results, at the expenses of network resources. Thus, a performance impact under information exchange constraint is proposed, as function of a T_I time slot and the average information size of N_I bits for each slot. By adjusting the quantity of N_I and the interval T_I, different network conditions can be created. For instance, by increasing N_I and decreasing T_I a better policy can be learnt, but that will in turn increase network overheads.

Although the information exchange would be decreased for less overhead, the complexity of data processing at the nodes would still be high. Hence, a function approximation with much smaller state space is preferred to the exact value-function. Then, the nodes of the network exchange these approximate value functions. The approximation reduces the details of the network overview, favouring certain features.

Finally, another improvement can simplify the complexity of the system, lowering the frequency of the approximate value function exchange. Since the difficulty of synchronization between the information reception and the moment an action is taken, the value-function is updated by all the rewards received during a T_I time slot. The consequence is that the TD learning has non-uniform $n(t)$-step temporal difference, which depends on the time a feedback from other nodes is received in relation to the state visited by a node.

In the results section, the average goodput per stage using the distributed actor-critic learning outperforms the centralized one and approaches the optimal policy. Then the AC learning has been evaluated with different n-step TD, showing the advantage of using a lower feedback frequency. When T_I is equal to 20 ms, the case with the least information feedback frequency, the average reward, without accounting for the information overhead, is higher than in the other cases (i.e., $T_I = 1, 10$). Also in the case accounting for the information overhead, its average goodput is higher than in the other two cases but with higher difference.

In [21], the idea of pairing two complementary radios with different frequency bands and maximum ranges/transmission power intensities is developed in order to provide more diversity. In this way, the sensor nodes, applying Q-learning, are able to determine at every event the proper radio and range calibration, aiming to minimize the energy expenses at variable interference without sacrificing much the packet reception. Indeed, this approach saves up to 52% of energy compared to single radio systems, handling node mobility.

9 Summary and Conclusions

Achieving dependable WSNs with low energy consumption and QoS guarantee is one of the biggest challenges in IoT. For many years, problems concerning battery depletion have been faced with classic solutions, using theoretical models and exploiting information and feedback from the network. In recent years, there has been an increasing interest in Machine Learning algorithms applied to WSN. The main goal of this chapter is to find similarities and differences, strength and weaknesses, among Transmission Power Control protocols. A classification of noteworthy protocols has been introduced, distinguishing two categories: deterministic and cognitive solutions. In the deterministic approach, the protocols are divided in proactive and reactive, where in the first case, transmission power levels are chosen besides an event may happen and so are determined prior to a packet transmission. In the second case, the transmission power is adjusted *a posteriori*, after the transmission of a packet, receiving feedback from either the network or the environment. The cognitive protocols are classified based on the self-learning methods: Swarm Intelligence, Fuzzy Logic and Reinforcement Learning.

Whereas many protocols target energy efficiency, fewer contributions have been proposed for QoS management, which is still an uncovered issue that requires more attention. This review suggests that greater efforts are needed in research to obtain reliable WSNs. While in fact TPC is a promising solution for both energy efficiency and QoS support, only few articles explore this potential. This seems the most promising research direction for those looking at methods that combine energy with spectrum efficiency.

Acknowledgements This work was jointly supported by NXP Semiconductors and Eindhoven University of Technology (IMPULS program) and by Horizon 2020 INTER-IoT project (grant 687283).

References

1. Al-Karaki, J.N., Kamal, A.E.: Routing techniques in wireless sensor networks: a survey. IEEE Wireless Commun. (2004). doi:10.1109/MWC.2004.1368893
2. Bosman, H.H.W.J., Liotta, A., Iacca, G., Wörtche, H.J.: Online extreme learning on fixed-point sensor networks. In: IEEE International Conference on Data Mining Workshops (2013). doi:10.1109/ICDMW.2013.74

3. Bosman, H.H.W.J., Iacca, G., Wörtche, H.J., Liotta, A.: Online fusion of incremental learning for wireless sensor networks. In: IEEE International Conference on Data Mining Workshop (2014). doi:10.1109/ICDMW.2014.79
4. Bosman, H.H.W.J., Iacca, G., Tejada, A., Wörtche, H.J., Liotta, A.: Ensembles of incremental learners to detect anomalies in ad hoc sensor networks. Ad Hoc Netw. (2015). doi:10.1016/j.adhoc.2015.07.013
5. Bosman, H.H.W.J., Iacca, G., Tejada, A., Wörtche, H.J., Liotta, A.: Spatial anomaly detection in sensor networks using neighborhood information. Inf. Fusion (2017). doi:10.1016/j.inffus.2016.04.007
6. Chen, B., Jamieson, K., Balakrishnan, H., Morris, R.: Span: an energy-efficient coordination algorithm for topology maintenance in ad hoc wireless networks. Wirel. Netw. (2002). doi:10.1023/A:1016542229220
7. Chen, D., Varshney, P.K.: QoS support in wireless sensor networks: a survey. In: International Conference on Wireless Networks (2004)
8. Chincoli, M., Bacchiani, C., Syed, A.A., Exarchakos, G., Liotta, A.: Interference mitigation through adaptive power control in wireless sensor networks. In: IEEE International Conference on Systems, Man, and Cybernetics (SMC) (2015). doi:10.1109/SMC.2015.232
9. Chincoli, M., Syed, A.A., Exarchakos, G., Liotta, A.: Power control in wireless sensor networks with variable interference. Hindawi Mob. Inf. Syst. (2016). doi:10.1155/2016/3592581
10. Collotta, M., Scat, G., Tirrito, S., Ferrero, R., Rebaudengo, M.: A parallel fuzzy scheme to improve power consumption management in Wireless Sensor Networks. In: Proceedings of the 2014 IEEE Emerging Technology and Factory Automation (ETFA) (2014). doi:10.1109/ETFA.2014.7005363
11. Das, B., Bharghavan, V.: Routing in ad-hoc networks using minimum connected dominating sets. In: IEEE International Conference on Communications (2014). doi:10.1109/ICC.1997.605303
12. Du, P., Roussos, G.: Adaptive communication techniques for the internet of things. J. Sens. Actuator Netw. (2013). doi:10.3390/jsan2010122
13. Exarchakos, G., Oztelcan, I., Sarakiotis, D., Liotta, A.: plexi: adaptive re-scheduling web-service of time synchronized low-power wireless networks. J. Netw. Comput. Appl. (2016). doi:10.1016/j.jnca.2016.06.013
14. Fortino, G., Trunfio, P.: Internet of Things Based on Smart Objects, Technology, Middleware and Applications. Springer, Berlin (2014). doi:10.1007/978-3-319-00491-4
15. Fortino, G., Gravina, R., Li, W., Hassan, M.M., Liotta, A.: Enhancing internet and distributed computing systems with wireless sensor networks. Int. J. Distrib. Sens. Netw. (2015). doi:10.1155/2015/564695
16. Galzarano, S., Fortino, G., Liotta, A.: Embedded self-healing layer for detecting and recovering sensor faults in body sensor networks. In: IEEE International Conference on Systems, Man, and Cybernetics (2012). doi:10.1109/ICSMC.2012.6378098
17. Galzarano, S., Liotta, A., Fortino, G.: QL-MAC: A Q-Learning Based MAC for Wireless Sensor Networks. Springer, Berlin (2013)
18. Galzarano, S., Fortino, G., Liotta, A.: A learning-based MAC for energy efficient wireless sensor networks. In: Internet and Distributed Computing Systems. Springer (2014)
19. Geelen, D., van Kempen, G., van Hoogstraten, F., Liotta, A.: A wireless mesh communication protocol for smart-metering. In: International Conference on Computing, Networking and Communications (ICNC) (2012). doi:10.1109/ICCNC.2012.6167440
20. Grace, D., Zhang, H.: Cognitive Communications: Distributed Artificial Intelligence (DAI), Regulatory Policy and Economics, Implementation. Wiley, New York (2012)
21. Gummeson, J., Ganesan, D., Corner, M.D., Shenoy, P.: An adaptive link layer for heterogeneous multi-radio mobile sensor networks. IEEE J. Sel. Areas Commun. (2010). doi:10.1109/JSAC.2010.100914
22. Hackmann, G., Chipara, O., Lu, C.: Robust topology control for indoor wireless sensor networks. In: Proceedings of the 6th ACM Conference on Embedded Network Sensor Systems (2008)

23. Kanaris, L., Kokkinis, A., Raspopoulos, M., Liotta, A., Stavrou, S.: Improving RSS fingerprint-based localization using directional antennas. In: The 8th European Conference on Antennas and Propagation (2014). doi:10.1109/EuCAP.2014.6902090
24. Kanaris, L., Kokkinis, A., Fortino, G., Liotta, A., Stavrou, S.: Sample Size Determination Algorithm for fingerprint-based indoor localization systems. Comput. Netw. (2016). doi:10.1016/j.comnet.2015.12.015
25. Kim, J., Chang, S., Kwon, Y.: ODTPC: on-demand transmission power control for wireless sensor networks. In: IEEE International Conference on Information Networking (2008)
26. Kobbane, A., Koulali, M.A., Tembine, H., Koutbi, M.E., Ben-othman, J.: Dynamic power control with energy constraint for Multimedia Wireless Sensor Networks. In: IEEE International Conference on Communications (ICC) (2012). doi:10.1109/ICC.2012.6363971
27. Kokkinis, A., Raspopoulos, M., Kanaris, L., Liotta, A., Stavrou, S.: Map-aided fingerprint-based indoor positioning. In: IEEE 24th Annual International Symposium on Personal, Indoor, and Mobile Radio Communications (2013). doi:10.1109/PIMRC.2013.6666144
28. Kokkinis, A., Kanaris, L., Raspopoulos, M., Liotta, A., Stavrou, S.: Optimizing route prior knowledge for map-aided fingerprint-based positioning systems. In: The 8th European Conference on Antennas and Propagation (2014). doi:10.1109/EuCAP.2014.6902232
29. Kotian, R., Exarchakos, G., Mocanu, D.C., Liotta, A.: Predicting battery depletion of neighboring wireless sensor nodes. In: Algorithms and Architectures for Parallel Processing. Springer (2013). doi:10.1007/978-3-319-03889-6_32
30. Kotian, R., Exarchakos, G., Stavrou, S., Liotta, A.: Impact of transmission power control in multi-hop networks. Elsevier Future Gen. Comput. Syst. J. (2016)
31. Kulkarni, R.V., Forster, A., Venayagamoorthy, G.K.: Computational intelligence in wireless sensor networks: a survey. IEEE Commun. Surveys Tutor. (2011). doi:10.1109/SURV.2011.040310.00002
32. Kulkarni, R.V., Venayagamoorthy, G.K.: Particle swarm optimization in wireless-sensor networks: a brief survey. IEEE Trans. Syst. Man, Cybern. Part C (Appl. Rev.) (2011). doi:10.1109/TSMCC.2010.2054080
33. Lin, Z., van der Schaar, M.: Autonomic and distributed joint routing and power control for delay-sensitive applications in multi-hop wireless networks. IEEE Trans. Wirel. Commun. (2011). doi:10.1109/TWC.2010.111910.091238
34. Liotta, A., Geelen, D., van Kempen, G., van Hoogstraten, F.: A survey on networks for smart-metering systems. Int. J. Pervasive Comput. Commun. (2012). doi:10.1108/17427371211221072
35. Liotta, A.: Farewell to deterministic networks. In: IEEE 19th Symposium on Communications and Vehicular Technology in the Benelux (SCVT) (2012). doi:10.1109/SCVT.2012.6399413
36. Liotta, A.: The cognitive NET is coming. IEEE Spectr. (2013). doi:10.1109/MSPEC.2013.6565557
37. Lin, S., Zhang, J., Zhou, G., Gu, L., He, T., Stankovic, J.A.: ATPC: adaptive transmission power control for wireless sensor networks. In: ACM Proceedings of the 4th International Conference on Embedded Networked Sensor Systems (2006)
38. Liu, Z., Elhanany, I.: RL-MAC: a reinforcement learning based MAC protocol for wireless sensor networks. Int. J. Sens. Netw. (2006)
39. Lucas Martínez, N., Martínez Ortega, J., Hernández Díaz, V., del Toro Matamoros, R.M.: Communication range dynamics and performance analysis for a self-adaptive transmission power controller. Sensors (2016). doi:10.3390/s16050684
40. Masood, M.M.Y., Ahmed, G., Khan, N.M.: A Kalman filter based adaptive on demand transmission power control (AODTPC) algorithm for wireless sensor networks. In: IEEE International Conference on Emerging Technologies (ICET) (2012)
41. Mocanu, D.C., Exarchakos, G., Liotta, A.: Node centrality awareness via swarming effects. In: IEEE International Conference on Systems, Man and Cybernetics (SMC) (2014). doi:10.1109/SMC.2014.6973878
42. Pandana, C., Liu, K.J.R.: Near-optimal reinforcement learning framework for energy-aware sensor communications. IEEE J. Sel. Areas Commun (2005). doi:10.1109/JSAC.2005.843547

43. Pantazis, N.A., Vergados, D.D.: A survey on power control issues in wireless sensor networks. IEEE Commun. Surveys Tutor. (2007)
44. Rappaport, T.S.: Wireless Communications: Principles and Practice. Prentice Hall, Englewood Cliffs (2001)
45. Saleem, M., Di Caro, G.A., Farooq, M.: Swarm intelligence based routing protocol for wireless sensor networks: survey and future directions. Inf. Sci. (2011). doi:10.1016/j.ins.2010.07.005
46. Sanchez, E.R., Montrucchio, B., Murillo, L.M., Rebaudengo, M.: Adaptive Fuzzy-MAC for power reduction in wireless sensor networks. In: 4th IFIP International Conference on New Technologies, Mobility and Security (NTMS) (2011). doi:10.1109/NTMS.2011.5720629
47. Sheng, Z., Yang, S., Yu, Y., Vasilakos, A.V., Mccann, J.A., Leung, K.K.: A survey on the IETF protocol suite for the internet of things: standards, challenges, and opportunities. IEEE Wirel. Commun. (2013). doi:10.1109/MWC.2013.6704479
48. Singh, S., Raghavendra, C.S.: PAMAS–Power aware multi-access protocol with signalling for ad hoc networks. SIGCOMM Comput. Commun. Rev. (1998). doi:10.1145/293927.293928
49. Son, D., Krishnamachari, B., Heidemann, J.: Experimental study of the effects of transmission power control and blacklisting in wireless sensor networks. In: IEEE Sensor and Ad Hoc Communications and Networks (2004)
50. Sutton, R.S., Barto, A.G.: Introduction to Reinforcement Learning. MIT Press, Cambridge (1998)
51. Van Dam, T., Langendoen, K.: An adaptive energy-efficient MAC protocol for wireless sensor networks. In: Proceedings of the 1st International Conference on Embedded Networked Sensor Systems. ACM (2003)
52. Wimalajeewa, T., Jayaweera, S.K.: Optimal power scheduling for correlated data fusion in wireless sensor networks via constrained PSO. IEEE Trans. Wirel. Commun. (2008). doi:10.1109/TWC.2008.070386
53. Xu, Y., Heidemann, J., Estrin, D.: Geography-informed energy conservation for ad hoc routing. In: ACM Proceedings of the 7th Annual International Conference on Mobile Computing and Networking (2001). doi:10.1145/381677.381685
54. Yau, K.A., Goh, H.G., Chieng, D., Kwong, K.H.: Application of reinforcement learning to wireless sensor networks: models and algorithms. Computing (2015). doi:10.1007/s00607-014-0438-1
55. Ye, W., Heidemann, J., Estrin, D.: Medium access control with coordinated adaptive sleeping for wireless sensor networks. IEEE/ACM Trans. Netw. (2004). doi:10.1109/TNET.2004.828953
56. Zhao, Q., Grace, D., Clarke, T.: Transfer learning and cooperation management: balancing the quality of service and information exchange overhead in cognitive radio networks. Trans. Emerging Tel. Technol. (2015). doi:10.1002/ett.2913
57. Zheng, T., Radhakrishnan, S., Sarangan, V.: PMAC: an adaptive energy-efficient MAC protocol for wireless sensor networks. In: 19th IEEE International Parallel and Distributed Processing Symposium (2005). doi:10.1109/IPDPS.2005.344

Devices Can Be Secure and Easy to Install on the Internet of Things

Roger D. Chamberlain, Mike Chambers, Darren Greenwalt, Brett Steinbrueck and Todd Steinbrueck

Abstract One of the major issues that must be addressed in the emerging Internet of Things (IoT) is balancing the needs of security and reasonable installation and maintenance efforts. Security is crucial, as evidenced by the fact that IoT devices are frequent targets of attack. However, if the security infrastructure is not relatively easy to use, it will ultimately be compromised by users who are unwilling (or insufficiently motivated) to deal with the complexity of ensuring security. This paper describes the industrial deployment experience of the EZConnect™ security infrastructure implemented by BECS Technology, Inc., a firm that provides water chemistry monitoring and control equipment to the aquatics market.

1 Introduction

Firewalls are a crucial element in modern cyber-security deployments. However, they are also a substantial impediment to integrating devices in the Internet of Things (IoT) [13]. When IoT devices are attached to a local-area network, it is frequently the case that any attempt to contact these devices remotely (from outside the local-area network) is blocked by a firewall. Authorized remote access requires explicit intervention in the security infrastructure meant to protect the local-area network

R.D. Chamberlain (✉)
Dept. of Computer Science and Engineering, Washington University in St. Louis, St. Louis, MO, USA
e-mail: roger@wustl.edu

R.D. Chamberlain · M. Chambers · D. Greenwalt · B. Steinbrueck · T. Steinbrueck
BECS Technology, Inc., St. Louis, MO, USA
e-mail: mike_c@becs.com

D. Greenwalt
e-mail: darren@becs.com

B. Steinbrueck
e-mail: brett@becs.com

T. Steinbrueck
e-mail: todd@becs.com

R. Gravina et al. (eds.), *Integration, Interconnection, and Interoperability of IoT Systems*, Internet of Things, DOI 10.1007/978-3-319-61300-0_4

from attack. For field-area network installations, an even greater set of issues must be considered [30]. This paper describes the industrial deployment experience of the EZConnect™ security infrastructure implemented by BECS Technology, Inc., a firm that provides water chemistry monitoring and control equipment to the aquatics market. EZConnect allows for remote, authorized communication with devices installed behind firewalls without compromising local-area network security. In addition, it supports installation with minimal interaction required between the installation team and the organization's IT department.

Many approaches to enable remote access can compromise the security of the local-area network. Yet it is important not to compromise security, rather we must enhance it, since security threats are real [25]. A number of examples of security breaches are described below.

In 2001, an individual was convicted of hacking into a computerized waste water management system in Queensland, Australia, causing raw sewage to spill into local parks and rivers [34]. From 2003 to 2006, at least four cyber-attacks on water supply systems in the U.S. were reported to WaterISAC, an industry information sharing and analysis center. In one of these attacks, the attackers declared their presence with the message, "I enter in your server like you in Iraq." [12]. More recently, hackers infiltrated a water utility's control system and changed the levels of chemicals in the tap water treatment process [23]. The industry magazine *Industrial Waterworld* reports that, as a test of security vulnerability, three industrial control systems used in the water industry were connected to the Internet to see if they would be attacked [22]. It took about 18 hours. In the course of 28 days, the devices were attacked 39 times. "The attacks included modifying settings to change water pressure and stop the flow on a water pump."

In spite of these real threats, if a security infrastructure is overly burdensome, it will either not be used at all or will be diminished in effectiveness by lack of diligence on the part of the owners/operators and third-party support/service providers. To be truly effective, any approach to security must be paired with an approach to ease the burden on the user [15]. Hertzum et al. [17] assessed the intrinsic tensions between security and ease of use in an e-banking context, and concluded that ease of use limitations can directly contribute to decreased security. Schneider [33] cautions us all that security must not rely on unreasonable expectations about the actions of users. "We must stop trying to fix the user to achieve security."

We describe the industrial deployment experience of BECS Technology's approach for providing secure connectivity to installed embedded IoT devices in aquatics. BECS Technology is a manufacturer of monitoring and control equipment in the aquatics, agriculture, cable, and refrigeration markets. BECS Technology provides remote access capability to its devices for owners/operators from both desktop software and mobile apps for smartphones and tablets.

Marketed under the trade name EZConnect™, we describe an approach to supporting remote communication with IoT devices that satisfies the need for security yet balances that need with the equivalent need for ease of installation and maintenance. After discussing the prior security approaches that we previously suggested to our customers, we will describe EZConnect, what it is and how it works, as well as describe the security layers it embodies and properties that facilitate ease of use and maintenance. We will also comment on its acceptance in the marketplace.

The organization of the paper is as follows. Section 2 provides background on the context of our controllers within the IoT and describes related work. Section 3 talks about prior approaches to ensuring that IoT devices are secure, including the ease of use limitations that those approaches imply. Section 4 describes EZConnect, our approach to secure remote communications capability, and Sect. 5 articulates the layers of security built into EZConnect. Section 6 describes how EZConnect improves the user experience, and Sect. 7 documents its acceptance in the marketplace. Section 8 concludes and describes future work.

2 Background

2.1 Context Within the Internet of Things

BECS Technology's controllers are fairly typical devices in the Internet of Things (IoT) [13]. The controller itself monitors various aspects of water chemistry: pH, oxidation-reduction potential (ORP), free chlorine concentration, temperature, conductivity, turbidity, etc. Based on these readings, the controller takes various actions (feeding chemical, adjusting recirculation flow rate, etc.) to maintain the water chemistry within desired parameters. Alarm conditions trigger notifications to service personnel. Sensor values and actions are all logged internally, and these logs are frequently used when diagnosing the causes of alarms. Remote access to all of the above information is clearly to the benefit of the equipment owners/operators.

While the notion of IoT might be new, the fundamental capability to access controller information remotely is not. BECS Technology's controllers have supported remote communications for more than 2 decades. Early controllers used modems attached to the telephone network, today controllers support TCP/IP connectivity via the Internet.

Remote capabilities include viewing of current status, downloading of data logs, and configuration of the controller. Figure 1 illustrates a realtime view of water chemistry for a specific body of water as remotely displayed on a PC screen. Four readings are being shown: pH (at 7.1), ORP (at 750 mV), temperature (at 83 °F), and free chlorine (at 1.0 ppm). Also indicated are the set points and high and low alarm points for each reading as well as the control outputs (in this example control is based on pH and ORP). The two dials at the bottom show a pair of indices (Langelier Saturation Index and Ryznar Stability Index) which are indicators of the scaling properties of the water [28]. The panel on the left allows the user to navigate to different controllers (either at the same or a different physical location). The tabs at the top allow the user to access a menu tree that can examine and/or modify a multitude of parameters on the controller.

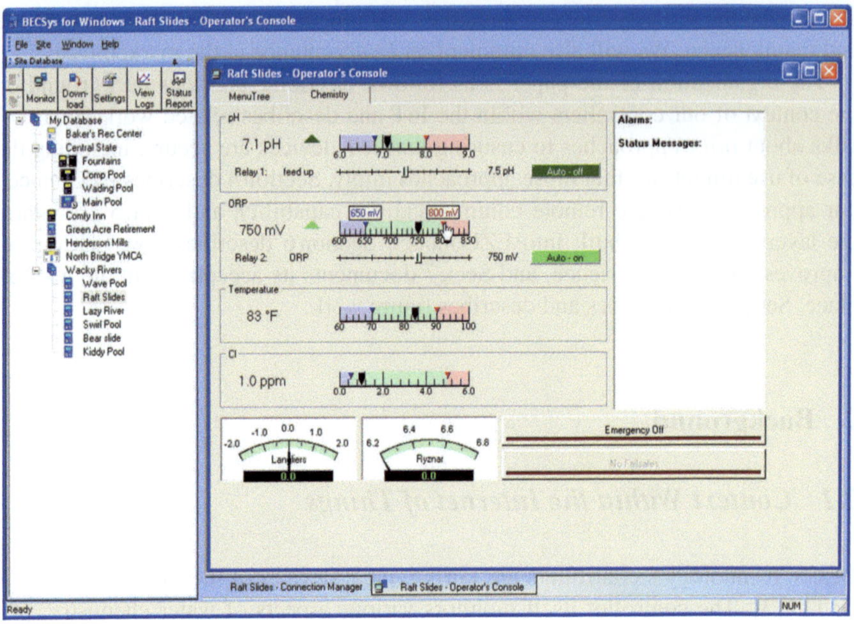

Fig. 1 Console display of controller at remote location

Figure 2 illustrates data logs collected over a 2 day period. The same four readings are plotted on the top graph, and the bottom graph gives indications of control actions, alarms, and other events. In the top curve, one can readily see disturbances in the pH at noon on both days. The additional window on the right shows the instantaneous values at the position of the cursor (the vertical line positioned by the user at 9am on the first day). As above, the leftmost panel supports navigation to different controllers.

To reinforce the notion that remote communications capability is nothing new, the data logs shown are from a time period more than a decade ago (in 2006). While the two figures show images from a desktop PC screen, modern remote communications capability is also supported via apps that run on smartphones and tablets.

In addition to diagnosing the root causes of errors in the water chemistry, the historical logs also enable the tracking of parameter changes by operators as well as support the demonstration and documentation of regulatory compliance.

In an ideal world (one in which security were not an issue), the connection between applications (either PC-based or mobile apps) and the controller would appear as shown in Fig. 3. The controller would have an IP address on the network, and applications could directly open a connection to the controller. Essentially, this was the topology that was common in the days of telephone modems, with the IP address replaced by a telephone number and the Internet cloud replaced by the telephone network.

Modern security concerns, however, preclude such a network topology today. Controllers are connected to a local-area network, and invariably that local-area

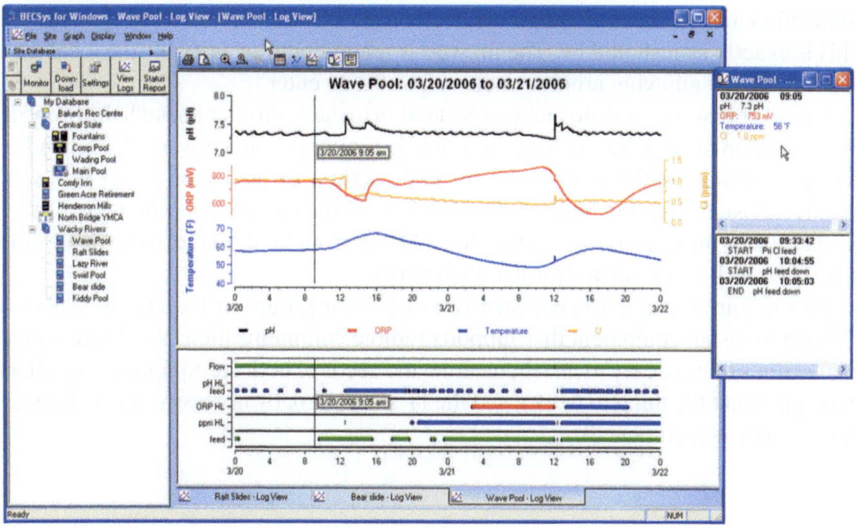

Fig. 2 Plot of controller data logs

Fig. 3 Idealized device interaction

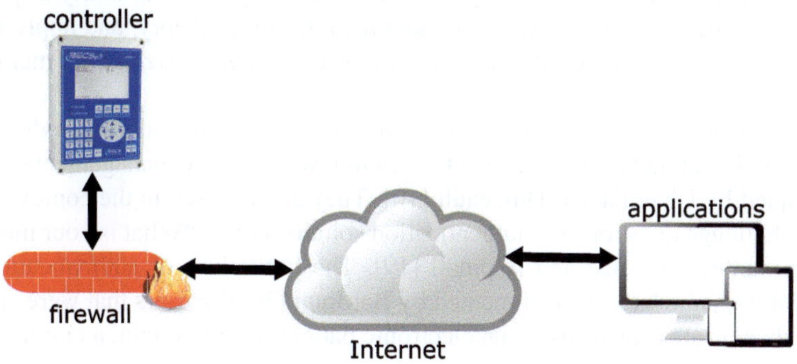

Fig. 4 Realistic device interaction

network is protected by a firewall. This more realistic connection topology is shown in Fig. 4.

The implications of the controller being installed on a local-area network behind a firewall are significant. Under normal circumstances, any attempt to initiate a connection to the controller from outside the firewall will be blocked by the firewall. This is exactly as it should be. The firewall is maintaining security for the local-area network by not allowing arbitrary network traffic to enter.

It is, of course, possible to make a connection to the controller through the firewall. Two common approaches include the use of a virtual private network (VPN) and/or port forwarding on the part of the firewall. Setting up either of these, however, requires close interaction between the installers of the equipment (who typically are *not* familiar with networking) and the local IT department personnel (who one would not expect to be familiar with water chemistry).

This required interaction between two dissimilar groups of individuals makes it difficult to install equipment that supports remote communication. We desire secure connections that are easy to install, meeting the needs of both the equipment installers (who just want the thing to work), and the IT staff (who, by necessity, have to ensure that security is maintained).

2.2 Related Work

There has been much study into how usability and security interact. Cranor and Garfinkel [8], in their introduction to a journal special issue dedicated to the subject, acknowledge that the conventional wisdom is that these two issues can be at odds with one another. They contend, however, that this notion goes against "common sense and more than 30 years' experience in building secure systems."

In the above special issue, Yee [35] asserts that conflicts between usability and security goals can be avoided if considered appropriately during the design process. Braz et al. [6] introduce the Security Usability Symmetry (SUS) usability inspection method as a design model, exploiting automata machines theory, and apply it to a banking software use case. SUS is a variant of the *Heuristic Evaluation* method of Nielsen [29].

The tradeoffs between usability and security in personal firewalls are evaluated by Herzog and Shahmehri [18]. Their interaction in web identity management systems is critiqued by Dhamija and Dusseault [10]. They are assessed in the context of the use of challenge questions for authentication solutions (e.g., "What is your mother's maiden name?") by Just and Aspinall [19]. The limitations of authentication via login for nomadic users is documented by Bardram [3]. The users that were studied are medical professionals using an electronic patient record system, a circumstance where compromised security is of particular societal concern.

The importance of social issues was investigated by Gaw et al. [14], who investigated the use (or lack of use) of encrypted email. In addition to the relevant technical factors, such as usability, they found that social factors were also quite relevant in users' choices.

The above research, however, is primarily concerned with the experience of software users, not equipment installers. For devices on the Internet of Things, the "things" are frequently pieces of equipment, and while the users definitely include end-users that access the equipment via software, equipment installers are a distinct category of user whose needs must be considered.

In many ways, security for IoT devices cannot simply follow the approaches developed in other areas. Heer et al. [16] provide an overview of security challenges for IP-based IoT installations, including a security deployment model, security needs, and requirements. Riahi et al. [31] describe a new security paradigm for IoT, proposing a systemic approach that explores the roles of new actors (relative to traditional contexts) and their interactions. Zhang et al. [36] articulate ongoing challenges and research opportunities in the area of IoT security, and Cui and Stolfo [9] and Costin et al. [7] have identified significant numbers (hundreds of thousands) of IoT devices with exposed vulnerabilities.

3 Prior Practice

Prior to the development of EZConnect, the users of controllers manufactured by BECS Technology were required to either use a VPN or have their IT department enable port forwarding in their firewall.

For many, the standard-of-practice for remote secure access to a local-area network that is protected behind a firewall is the use of a VPN [27], or virtual private network. It has the advantage of common use, so that the available solutions have enough history that they can be reasonably trusted. In addition, it is a solution that IT departments are very familiar with, providing a certain level of institutional comfort.

However, truly secure VPN solutions can have significant overhead both in terms of setup, maintenance, and individual use. First, VPNs are managed and maintained by the IT department, so by definition they require the active involvement of the IT department, both for initial setup and for any changes (e.g., account changes due to personnel transitions, etc.). Second, it is difficult (often against organizational policy) in many cases to provide a VPN account to those who are not employees of the organization. However, as we will describe below in Sect. 6, temporary remote access to the equipment (by third parties for diagnostic purposes) is often desired by equipment managers. Third, a VPN account gives access to the local-area network as a whole, not just an individual piece of equipment on the network. Fourth, many VPN solutions require a fob or other physical authentication device that can have quantity or institutional policy limitations.

The other approach commonly used to provide remote access through a firewall into a local-area network is via port forwarding [2]. Here, an association is made within the firewall configuration that enables an attempted connection outside the network to be completed inside the network.

This approach also requires the involvement of the IT department; however, it gives the misleading impression that the IT department's involvement is substantially lower than with the VPN solution. The idea is that once the port forwarding is setup, IT need no longer be involved. In fact, unless the port forwarding associations are updated (e.g., when personnel change), it is not truly secure, and when the associations are changed, both IP addresses and ports used to remotely access equipment must be updated. This approach also has the downside that users must now use a different IP address/port combination when accessing the equipment from within the local-area network versus when accessing the equipment remotely.

In effect, both of these approaches have the significant disadvantage that they each heavily involve IT department resources (i.e., they take time to setup and maintain) and also raise legitimate security concerns on the part of IT managers. This is particularly true for the port forwarding approach, which can become a security hole if not carefully monitored (see, for example, the description of the 'port forward' exploit described by Ammann et al. [1]). From a usability point of view, the need to distribute (and maintain) IP addresses is a clear negative.

There is an additional downside for approaches that have substantial interaction between IT department personnel and equipment managers and installers. They are two distinct communities that have different technical language, background experiences, and missions within the organization. On the IT side, the vocabulary includes terms and acronyms such as "IP address," "DHCP," "port forward," and "VPN." This is all very relevant to effective and secure communication that is crossing a firewall, but is completely alien to users and installers of water treatment equipment. On the installer side, the vocabulary terms include "pH," "oxidation-reduction potential," "alkalinity," and "turbidity." These are concepts that are essential to clean water, but just as unfamiliar to IT personnel as "DNS" is to them.

What the above implies is that simply expecting each party to the conversation to readily be comfortable with any solution is naive if it is predicated on one group having to become familiar with the other's language and expertise. We seek a technical solution that: (1) minimizes the required interaction, to reduce the effort needed to get the job done, and (2) provides an effective, secure communications capability that is in the comfort zone of all involved.

4 EZConnect

With the introduction of EZConnect, water treatment equipment manufactured by BECS Technology supports effective, secure communications while addressing the concerns expressed above. In this section, we will describe the logistics of the system, the essentials of how it works.

The typical configuration of an EZConnect installation is depicted in Fig. 5. The controller on the left is connected via a local-area network and sits behind a firewall. Applications (either desktop programs or mobile apps) on the right wish to communicate with the controller; however, attempts to connect directly to the controller via the Internet are normally blocked by the firewall.

Rather than have any device outside the firewall attempt a direct connection, the approach used by EZConnect is to have the controller, which is inside the firewall, make an outbound connection to the EZConnect server. This is shown in Fig. 6. Whenever the EZConnect option is enabled on a controller, the controller will

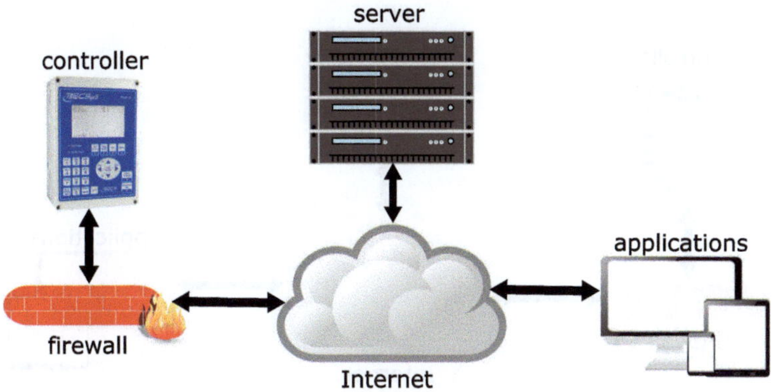

Fig. 5 EZConnect system diagram

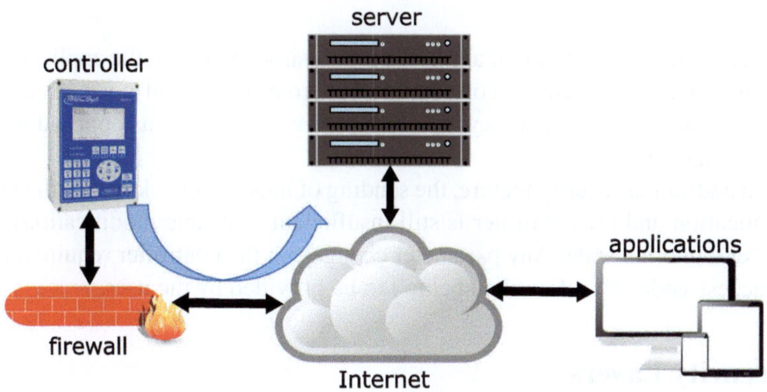

Fig. 6 Controller connects to server

automatically contact the server and register the controller. The controller then maintains an open socket on the server, which will subsequently be used for authenticated remote connections. The server will then wait for an application to request a connection to the controller.

This redirection of the initial contact is essential for the goals of the system to be realized. First, outbound connections are much less problematic for security-sensitive IT departments than are inbound connections. Second, it is not necessary for the controller's IP address to be known by any entity outside of the local-area network.

When an application wishes to connect to a controller, rather than attempting to connect to the controller directly, contact is made with the EZConnect server, as illustrated in Fig. 7. This eliminates the need for a dedicated access path through the site's router/firewall. The application must provide both the serial number of the controller as well as a valid authentication code (described below), upon which communication is established by the server, forwarding messages between the application and the

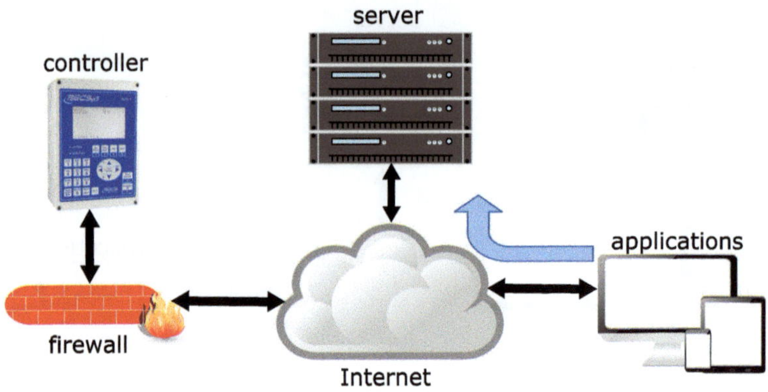

Fig. 7 Applications connect to server

controller. At this point, individual messages forwarded between the application and the controller are checked for correctness, but from a practical point of view, the applications are unaware that they are connected via the server as opposed to being directly connected.

As an additional security feature, the sending of messages back and forth between the application and the controller is still insufficient to enable modifications to the controller's internal state. Any parameter changes on the controller require an additional access code (also described below) to be provided by the user.

5 Security Layers

Current best practices do not rely on an individual "silver bullet" to provide security, but instead rely on a layered approach, in which one component of the security infrastructure being compromised does not compromise the entire system. In the EZConnect system, there are a number of layers of security, each of which will be described below, along with some comments on the tradeoffs that they imply.

1. **TLS Encryption** – All communications between the controller(s), the server(s), and the application software are encrypted with the industry standard TLS (Transport Layer Security) cryptographic protocol [11].

   In addition to encrypting the data sent over the network, TLS also ensures that the server is the legitimate server and not an impostor with potentially malicious intent. TLS is used extensively to secure network communications in a number of areas, including e-commerce, web browsing, email, instant messaging, voice-over-IP (VoIP), etc.

2. **Proprietary Communication Protocols and Applications** – The protocol used to communicate with the controllers has intentionally limited semantic capability.

It only supports the download of data logs and reading/writing of parameters stored on the controller that deal with the controller's function as an aquatic controller. The protocol does not include commands to interact with the local network or embedded operating system, and cannot be used to upload software (either to the controller itself or to any other device on the network).

In addition to the semantic limitations, the protocol itself is unpublished and only supported by applications provided by BECS Technology. This has the effect of minimizing the attack surface exposed to potential threats [26], especially relative to systems that rely on protocols supported by any web browser (e.g., HTTP and HTML).

Maintaining this confidentiality has similar issues as maintaining any trade secret, once disclosed it is no longer a secret, and it does not encourage interoperability. The interoperability interest we have seen so far in the marketplace has been focused on enabling information about water chemistry to be available to wider building management systems. We plan to support this interaction through the EZConnect server, or through Modbus, but not directly via the controller's proprietary protocol.

The semantic constraints are definitely the strongest aspect of this security layer. Without the ability to specify malicious activity, the link between the server and the controller is demonstrably more difficult to exploit.

3. **Controller Access Codes** – Every controller has 3 levels of access code (password) protection: Operator, Manager, and Rep (a manufacturer's representative is typically responsible for installation and setup). Parameter changes on the controller can only be performed after the user has been granted one of these access levels, which are enforced for both local (controller front-panel user interface) and remote changes.

   Operators are allowed lowest-level access (e.g., establishing setpoints, recalibration of sensors, reset alarms, etc.). Managers have access to additional controller configuration options (e.g., default front-panel display) and establishing Operator accounts. Reps have full access to detailed control parameters (e.g., time-based proportional control setup) and various installation verification tools (e.g., relay overrides, etc.).

   The access codes are controlled and managed by the owner/operator of the controller. As such, they embody all the pros and cons of user control (e.g., users generally pick codes based on ease of remembering rather than difficulty in breaking). Currently, access codes only include numeric characters (which are unfortunately vulnerable to brute force attack), since the front panel of the controller does not support alphanumeric input; however, this is slated to be expanded to full alphanumeric support in the next generation of the controller line.

4. **EZConnect Authentication Codes** – In addition to the access codes described above, each controller has a list of authentication codes, which identify users that have been authorized to access that controller remotely. Authentication codes are 8-character alphanumeric values generated at random by the controller. Generating, changing, or deleting them requires either Manager or Rep access, persons with Operator access are not able to view or modify them. They are entered into the application that is requesting remote connectivity; however, they are never visible in the application (i.e., after validation, the applications display "validated" rather than show the authentication code).

   Since the controller creates randomly generated authentication codes, they are less susceptible to list-based attacks than user-generated passwords.

5. **Physical ROM Program Store** – The controllers are multi-processor systems, and the processor core that performs the actual control functions is a dedicated chip that reads its program only from a physical read-only memory (ROM). The only way to alter the code executed by this processor is to physically replace the program memory chip.

   While this design has clear implications for supporting the controllers (e.g., requiring physical access to provide software updates), it makes it that much harder for those with malicious intent to subvert the fundamental control function by replacing the controller code.

   The processor cores executing other functions are using an embedded Linux kernel that has state-of-the-art defenses integrated (e.g., see [20] for many of the techniques utilized). For example, the only open port is the one that is accessed by the proprietary protocol, and messages sent to that port that are malformed in any way do not receive any response.

One way in which the limited semantics of the proprietary interface supports secure operations is that the controller does not support software alterations from the network. We have already mentioned that the processor responsible for control operations requires a chip change to alter the software. In addition, we only support software updates for the remaining processor cores through a USB port on the controller.

Clearly, the overall security is also a function of the servers that provide connectivity to the controllers. The security of these servers is beyond the scope of this paper; however, we note that the above implies that even a server breach does not allow access to the local-area network on site. There are a number of resources that provide best practices for server security (e.g., see [5, 21, 32]).

The above security layers support the ease of installation goals as described next.

6 Ease of Installation and Maintenance

One of the primary motivations for the development of the EZConnect system was to ease the installation effort required on the part of both equipment managers and their colleagues in the IT department. Rather than asking the IT department to alter firewall settings (to support port forwarding) or setup a VPN, all that is required is that the firewall support outbound connections, which are commonly supported in modern deployments. In the event that outbound connections are disabled by default, it is sufficient for the IT department to enable outbound connections on only one port, a request that has only been denied by one customer's IT department to date (and this customer in not interested, by corporate policy, in supporting remote communications to controllers installed at its facilities).

The actual installation is almost completely plug and play. At startup time, the EZConnect option must be enabled (by default it is disabled) on the controller itself. The controller then initiates the connection to the server as described in Sect. 4. Once this connection is established, users can then connect (virtually) to the controller through the server.

This ease of installation should be contrasted with the steps necessary to enable remote access using either of the previously described methods. Installers needed to interact with the site's IT department, attempting to communicate using terms that are completely unfamiliar to them (e.g., IP address, subnet mask, etc.). Even simple transcription mistakes (which are made more likely by the unfamiliarity of the terminology) cause the system to fail, triggering diagnostic effort on the part of both the IT and equipment installation teams. The EZConnect installation approach has already shown itself to be vastly superior in the field.

One feature of the EZConnect system that has proven to be very popular is the ability for equipment managers to temporarily allow remote access to service personnel. When the water chemistry is an undesirable state (e.g., some alarm condition), it is not unusual for equipment owners to contact service personnel from either the manufacturer (BECS Technology) or the manufacturer's representative to help diagnose and correct the issue.

Under normal circumstances, these service personnel do not (and should not) have access to the equipment (unless as part of a formal service agreement). However, it is straightforward for a manager to create a new remote authentication code and provide it (along with an access code if desired) to the service personnel. Once the service is complete, the access and authentication codes can be readily disabled.

In this way, service personnel can be granted remote access to the controller for diagnostic purposes and potentially for corrective action, without requiring the intervention of the IT department to either enable or subsequently disable the access.

Contrast this with the approach required when secure access is being provided by a VPN. First, the IT department must be involved in both creating a new account and disabling it when no longer needed. Second, the access provided to the service personnel is not only to the controller, but is instead access to the internal network of the organization. This requires a higher level of trust than that needed with EZConnect.

Third, remote connectivity requires that the IP address of the controller be known outside the organization, a circumstance that is avoided entirely with EZConnect.

A manager's ability to disable remote authentication codes also supports the effective management of access in the case of departing employees. If each authorized employee has a unique authentication code, one individual's departure need not impact the other employees' credentials.

7 Marketplace Acceptance

Since the availability of EZConnect, virtually all newly installed controllers have used it for remote connectivity. A very small fraction have retained the use of VPNs, however, all of those new installations are actually at existing sites (e.g., the addition of a new controller or the replacement of a pre-existing controller) and already had dealt with the overhead of setting up a VPN or enabling port forwarding on the part of the firewall.

When examining the history of service calls related to remote connectivity, two facts stand out. First, virtually all of the service calls are associated with initial setup, rather than recurring issues. Initial configuration of the controllers is a complicated process, and there is regular communication between the factory and the field during this time. While setting up EZConnect gets discussed during these calls, it is not often the trigger for a call in the first place. Once configured, very few service calls relate to remote connectivity. Once working, it just keeps working, rarely causing a service call due to the lack of ability to communicate.

Second, prior to the existence of EZConnect, it was a common occurrance for those receiving service calls to refer the issue to the IT department. This regularly entailed a lengthy discussion of definitions of unfamiliar words and guidance as to how to interact with the IT department, and frequently resulted in several separate conversations over a period of days before the issue was resolved. With EZConnect, the service folks are typically able to enable proper setup in one call, including a verification of proper operation by BECS Technology personnel ensuring that the controller has registered with the servier. In summary, followup calls have been virtually eliminated and the need to involve the IT department in the eventual solution has diminished dramatically.

Here are a few of the stories we are hearing from our distributors:

- For a school's natatorium installation, the installation team ran into the issue that the state government had required all public schools to close all the external ports on their on-site equipment. Allowing a secure VPN connection was explicitly disallowed by the regulatory authorities, and the options available to the installation team were effectively reduced to zero.

  With EZConnect, the application note that describes how EZConnect works [4] was delivered to the customer, who forwarded it to the appropriate regulatory authorities, and the problem was effectively solved. The actions taken by the

controller and remote applications were within the acceptable operating parameters of the regulators.
- At installation time at one location, the local-area network was not yet available, so the installers enabled EZConnect on the controller and proceeded to wait until the local-area network installation could happen. Two months later, the Ethernet cable was connected, and the customer's remote access was immediate. No additional site visits were required.
- At two locations, repeated interactions with the IT department were required in an attempt to get the VPN configured correctly. One location finally worked, with substantial frustration on the part of both the installers and the IT department. The second location never was able to reliably connect to the controller. After enabling EZConnect, no problems were experienced at either site, and remote communications are working perfectly.

Since EZConnect has been generally available, only one installation has been told "no" by its IT department. This was the result of a policy decision on the part of the organization that it did not want to allow remote communications of any form.

8 Conclusions and Future Work

At the time of writing, controllers that support EZConnect have been through beta testing in between 50 and 100 installations, and they are currently being manufactured and installed as a released product.

So far, only one IT department has denied the request to support outbound connections through the firewall. This is a significant change from earlier circumstances, in which a typical installation required substantial meetings (and negotiation) with the IT department before connectivity might be supported, and it was all too frequently ultimately denied.

The EZConnect system has the following benefits:

- Highly secure with multiple layers of security measures.
- Hassle-free setup and operation.
- IT department does not need to establish VPN or forward ports on router.
- No need to know or distribute the controller IP address.
- User of remote application need only know controller serial number and authentication code.
- Equipment manager has full control over remote access.

There are, however, a number of limitations to be acknowledged. The use of a proprietary language is only secure while the trade secret is maintained, and a more robust character set for access codes would be less vulnerable to brute force attack. The use of proprietary mechanisms also limits interoperability. It is worth noting that none of these limitations are newly introduced with EZConnect, and with further refinements to the infrastructure they can all be addressed within the context of EZConnect. All in all, it is a nice compromise between the requirements for security and the reality that if it is not easy to use it will not get used.

There are a number of directions we are investigating that will further improve the system. Probably the most impactful is the desire to incorporate active intrusion detection mechanisms into the existing layered security approach. Within the controller itself, we can investigate the ideas presented by Li et al. [24], in which a separate, trusted OS runs concurrently with the general purpose Linux kernel and is charged with monitoring for integrity violations. More generally, intrusion detection mechanisms that are deployed on the server(s) in the cloud have the unique benefit of knowing what controllers have connected to them and are therefore potential targets for attack. They are therefore subject to intrusion detection monitoring.

We believe that the approach used in EZConnect is a model for how to connect to devices on the IoT in a convenient and secure manner.

References

1. Ammann, P., Wijesekera, D., Kaushik, S.: Scalable, graph-based network vulnerability analysis. In: Proceedings of 9th Conference on Computer and Communications Security, pp. 217–224. ACM, New York, NY, USA (2002). doi:10.1145/586110.586140
2. Apvrille, A., Pourzandi, M.: Secure software development by example. IEEE Secur. Priv. **3**(4), 10–17 (2005). doi:10.1109/MSP.2005.103
3. Bardram, E.: The trouble with login: on usability and computer security in ubiquitous computing. Pers. Ubiquitous Comput. **9**(6), 357–367 (2005). doi:10.1007/s00779-005-0347-6
4. BECSys EZConnect™ Application Note. Technical Report ENG-6072-DOC, BECS Technology, Inc., St. Louis, MO, USA (2016)
5. Binnie, C.: Linux Server Security: Hack and Defend. Wiley Inc, Indianapolis, IN, USA (2016)
6. Braz, C., Seffah, A., M'Raihi, D.: Designing a trade-off between usability and security: a metrics based-model. In: Baranauskas, C., Palanque, P., Abascal, J., Barbosa S.D.J. (eds.) Proceedings of IFIP TC-13 11th International Human-Computer Interaction Conference, Part II, pp. 114–126. Springer, Berlin (2007). doi:10.1007/978-3-540-74800-7_9
7. Costin, A., Zaddach, J., Francillon, A., Balzarotti, D.: A large-scale analysis of the security of embedded firmwares. In: Proceedings of 23rd USENIX Security Symposium, pp. 95–110. USENIX Association (2014). https://www.usenix.org/conference/usenixsecurity14/technical-sessions/presentation/costin
8. Cranor, L.F., Garfinkel, S.: Guest editors' introduction: secure or usable? IEEE Secur. Priv. **2**(5), 16–18 (2004). doi:10.1109/MSP.2004.69
9. Cui, A., Stolfo, S.J.: Reflections on the engineering and operation of a large-scale embedded device vulnerability scanner. In: Proceedings of 1st Workshop on Building Analysis Datasets and Gathering Experience Returns for Security, pp. 8–18. ACM, New York, NY, USA (2011). doi:10.1145/1978672.1978674
10. Dhamija, R., Dusseault, L.: The seven flaws of identity management: usability and security challenges. IEEE Secur. Priv. **6**(2), 24–29 (2008). doi:10.1109/MSP.2008.49
11. Dierks, T.: The Transport Layer Security (TLS) Protocol Version 1.2 (2008). http://tools.ietf.org/pdf/rfc5246.pdf. Rfc5246
12. Esposito, R.: Hackers penetrate water system computers. ABC News (2006). http://blogs.abcnews.com/theblotter/2006/10/hackers_penetra.html
13. Fortino, G., Trunfio, P. (eds.): Internet of Things Based on Smart Objects. Springer, Berlin (2014). doi:10.1007/978-3-319-00491-4

14. Gaw, S., Felten, E.W., Fernandez-Kelly, P.: Secrecy, flagging, and paranoia: adoption criteria in encrypted email. In: Proceedings of SIGCHI Conference on Human Factors in Computing Systems, pp. 591–600. ACM, New York, NY, USA (2006). doi:10.1145/1124772.1124862
15. Gefen, D., Straub, D.W.: The relative importance of perceived ease of use in IS adoption: a study of e-commerce adoption. J. Assoc. Inf. Syst. **1**(1), 8 (2000). http://aisel.aisnet.org/jais/vol1/iss1/8
16. Heer, T., Garcia-Morchon, O., Hummen, R., Keoh, S.L., Kumar, S.S., Wehrle, K.: Security challenges in the IP-based internet of things. Wirel. Pers. Commun. **61**(3), 527–542 (2011). doi:10.1007/s11277-011-0385-5
17. Hertzum, M., Jørgensen, N., Nørgaard, M.: Usable security and e-banking: ease of use vis-a-vis security. Australasian J. Inf. Syst. **11**(2) (2004). doi:10.3127/ajis.v11i2.124
18. Herzog, A., Shahmehri, N.: Usability and security of personal firewalls. In: Venter, H., Eloff, M., Labuschagne, L., Eloff, J., von Solms, R. (eds.) New Approaches for Security, Privacy and Trust in Complex Environments: Proceedings of IFIP TC-11 22nd International Information Security Conference, pp. 37–48. Springer, Boston, MA, USA (2007). doi:10.1007/978-0-387-72367-9_4
19. Just, M., Aspinall, D.: Personal choice and challenge questions: a security and usability assessment. In: Proceedings of 5th Symposium on Usable Privacy and Security, pp. 8:1–8:11. ACM, New York, NY, USA (2009). doi:10.1145/1572532.1572543
20. Kleidermacher, D., Kleidermacher, M.: Embedded Systems Security: Practical Methods for Safe and Secure Software and Systems Development. Elsevier, Waltham, MA, USA (2012)
21. Krutz, R.L., Vines, R.D.: Cloud Security: A Comprehensive Guide to Secure Cloud Computing. Wiley Inc, Indianapolis, IN, USA (2010)
22. Laughlin, J.: Industrial control systems targeted by hackers. Ind. WaterWorld **13**(3) (2013). http://www.waterworld.com/articles/iww/print/volume-13/issue-3.html
23. Leyden, J.: Water treatment plant hacked, chemical mix changed for tap supplies. The Register (2016). http://www.theregister.co.uk/2016/03/24/water_utility_hacked
24. Li, N., Kinebuchi, Y., Nakajima, T.: Enhancing security of embedded Linux on a multi-core processor. In: IEEE 17th International Conference on Embedded and Real-Time Computing Systems and Applications, vol. 2, pp. 117–121. IEEE, New York (2011). doi:10.1109/RTCSA.2011.36
25. Luiijf, E.: SCADA Security Good Practices for the Drinking Water Sector. TNO, The Hague, TNO-DV p. C096 (2008). http://m.tno.nl/media/1538/tno-dv-2008-c096_web.pdf
26. Manadhata, P., Wing, J.: An attack surface metric. IEEE Trans. Softw. Eng. **37**(3), 371–386 (2011). doi:10.1109/TSE.2010.60
27. Medaglia, C., Serbanati, A.: An overview of privacy and security issues in the internet of things. In: Giusto, D., Iera, A., Morabito, G., Atzori, L. (eds.) The Internet of Things, pp. 389–395. Springer, New York (2010). doi:10.1007/978-1-4419-1674-7_38
28. Müller-Steinhagen, H., Branch, C.: Comparison of indices for the scaling and corrosion tendency of water. Canadian J. Chem. Eng. **66**(6), 1005–1007 (1988). doi:10.1002/cjce.5450660617
29. Nielsen, J.: Finding usability problems through heuristic evaluation. In: Proceedings of SIGCHI Conference on Human Factors in Computing Systems, pp. 373–380. ACM, New York, NY, USA (1992). doi:10.1145/142750.142834
30. Palensky, P., Sauter, T.: Security considerations for FAN-Internet connections. In: Proceedings of IEEE International Workshop on Factory Communication Systems, pp. 27–35. IEEE, New York (2000). doi:10.1109/WFCS.2000.882530
31. Riahi, A., Challal, Y., Natalizio, E., Chtourou, Z., Bouabdallah, A.: A systemic approach for IoT security. In: Proceedings of International Conference on Distributed Computing in Sensor Systems, pp. 351–355. IEEE, New York (2013). doi:10.1109/DCOSS.2013.78
32. Ristić, I.: Bulletproof SSL and TLS: Understanding and Deploying SSL/TLS and PKI to Secure Servers and Web Applications. Feisty Duck, London, UK (2015)
33. Schneier, B.: Stop trying to fix the user. IEEE Secur. Priv. **14**(5), 96–96 (2016). doi:10.1109/MSP.2016.101

34. Smith, T.: Hacker jailed for revenge sewage attacks. The Register (2001). http://www.theregister.co.uk/2001/10/31/hacker_jailed_for_revenge_sewage
35. Yee, K.P.: Aligning security and usability. IEEE Secur. Priv. **2**(5), 48–55 (2004). doi:10.1109/MSP.2004.64
36. Zhang, Z.K., Cho, M.C.Y., Wang, C.W., Hsu, C.W., Chen, C.K., Shieh, S.: IoT security: Ongoing challenges and research opportunities. In: Proceedings of 7th International Conference on Service-Oriented Computing and Applications, pp. 230–234. IEEE, New York (2014). doi:10.1109/SOCA.2014.58

A Service-Based Approach for the Uniform Access of Wireless Sensor Networks and Custom Application Tasks Running on Sensor Nodes

Theodoros Fronimos, Manos Koutsoubelias, Spyros Lalis and Thomas Bartzanas

Abstract For reasons of convenience and interoperability, wireless sensor networks (WSNs) are typically accessed by remote client applications through well-defined service-oriented interfaces. In addition, significant flexibility can be gained by letting clients install and run custom, application-specific sensing and processing tasks directly on the nodes of the WSN. However, this raises the issue of how clients can interact with such tasks in parallel to the regular access of the WSN. We propose an approach for supporting both information flows simultaneously and in a uniform way, via an open web-based service interface that can be flexibly extended through appropriate XML descriptions which specify the format of the respective message exchanges. This allows clients to perform both the regular interactions with the WSN and the custom interactions with the application tasks that have been deployed on the sensor nodes, using a single and well-established interface technology. Moreover, the WSN gateway can perform the mapping between the client interface protocol and the WSN-level protocol for both regular and custom accesses, by using the same message parsing/conversion engine. The paper presents our design, describes a pro-

T. Fronimos · M. Koutsoubelias (✉) · S. Lalis · T. Bartzanas
Centre for Research and Technology Hellas, Institute for Research and Technology Thessaly, Volos, Greece
e-mail: emkouts@uth.gr

T. Fronimos
e-mail: frontheo@ireteth.certh.gr

T. Bartzanas
e-mail: bartzanas@ireteth.certh.gr

M. Koutsoubelias · S. Lalis
Electrical and Computer Engineering Department, University of Thessaly, Volos, Greece

S. Lalis
e-mail: lalis@uth.gr

R. Gravina et al. (eds.), *Integration, Interconnection, and Interoperability of IoT Systems*, Internet of Things, DOI 10.1007/978-3-319-61300-0_5

totype implementation, discusses an indicative application example, and evaluates the access overhead of our prototype and the benefits of deploying custom sensing tasks in the WSN.

1 Introduction

Wireless sensor networks (WSNs) are a key element of the modern computing landscape, supporting a broad range of monitoring and control applications. The clean and structured integration of WSNs into the Internet of Things (IoT) so that they can be exploited by third-party applications, emerges as a challenging issue [1, 5]. To this end, suitable interfaces are needed, which will allow different clients to obtain sensor measurements from selected or all nodes of the WSN at the desired rate.

At the same time, a large body of work proposes that WSNs be programmed in a more customized and dynamic way, by injecting application-specific code that runs directly on the nodes of the WSN. Such code can access the various on-board sensors directly, process the measurements locally, and send data towards external systems only when this is actually useful for the application. This can greatly reduce the amount of data that travels over the wireless links of the sensor network, which in turn can prolong the lifetime of energy-constrained nodes that are powered by batteries and energy scavenging mechanisms.

The ability to run custom code in the WSN clearly boosts flexibility, but also introduces a second dimension to the problem of transparent, structured and efficient access of the WSN. On the one hand, remote clients still need to issue queries and receive the data being produced by the sensor nodes, through a well-defined interface. The different parameters of such queries and the data that is produced as a response are known in advance, as they depend on the (fixed) hardware resources and firmware of the sensor nodes. On the other hand, the application-specific sensing and processing tasks that can be deployed in the WSN will typically support custom configuration/query requests and will produce custom data types. These are freely decided by the application developer, and thus cannot be part of the regular WSN interface. The question is how to support both the regular (static) and custom (dynamic) interface of the WSN in an open and uniform way.

As a solution to this problem, we propose a web-based interface through which clients can: (i) configure the sensor nodes of the WSN and query them to receive status information and measurements; (ii) deploy custom sensing and processing tasks on the sensor nodes; (iii) configure such tasks and receive the data that is produced by them. The syntax for both the regular and custom (task-specific) commands and data objects is defined using suitable XML descriptions. Clients simply need to format their requests to the WSN and parse the replies/data received from the WSN based on these descriptions. At the WSN gateway, these descriptions drive the necessary conversion between the web-based client protocol and the WSN protocol for the communication with the sensor nodes. At the sensor nodes, a runtime system handles the WSN-level protocol and takes over the dynamic installation and execution of custom application tasks.

The rest of the paper is structured as follows. Section 2 gives an overview of related work. Section 3 describes the high-level system architecture of our approach, Sect. 4 presents the interface of the WSN towards external clients, and Sect. 5 describes the programming model for custom application tasks. Section 6 discusses our current prototype implementation. Section 7 gives a concrete application example. Section 8 evaluates the access overhead of our system prototype, and illustrates the gains that can be achieved by deploying custom application tasks versus having long running queries in the WSN. Finally, Sect. 9 concludes the paper.

2 Background

Numerous publications address the issue of creating high-level abstractions for sensor networks from different points of view. Proposals such as TinyDB [23] and Cougar [30] follow a macro-programming approach by abstracting the whole WSN as a single database on which the user can perform complex queries. The main drawback is that the user has no control over in-network operations, as their optimization relies solely on the underlying middleware.

Contrary to macro-programming, virtual machine based approaches allow users to distribute processing among network nodes, in the form of mobile code. In middleware systems like Agilla [24] and Mate [21], mobile agents carrying user defined programs are injected or flooded throughout the sensor network. Programs are composed of special, low-level instructions (opcodes) following an assembly-like style. The overhead for performing application updates is much lower compared to methods that substitute the whole firmware image. However, application programming tends to be tedious and error prone.

Virtual Sensor Networks [18] elaborate on the concept of virtualization by adding support for the concurrent execution of multiple applications on a single node or subsets of nodes of a sensor network. In SenShare [20] the user is allowed to install, over the air, new applications without terminating the ones that were previously deployed. This is achieved through a runtime environment that implements a multi-threading execution model and provides an API for accessing system resources. SenShare focuses on the WSN architecture and does not deal with issues regarding interconnection with external systems.

To simplify WSN programming, TENET [25] introduces a set of predefined high-level operations, called tasklets, which are supported by the nodes of the WSN and can be executed on demand. The programmer can then compose application-specific tasks in a simple way, as a sequence of such operations. A task is encoded as a character string, sent to the sensor nodes for execution. In Servilla [13] heterogeneous nodes are pre-programmed with their own native services and a specification for each one of them is stored in a local registry. Application tasks use the declared services to access resources and are composed in a high level programming language. Their successful execution relies on whether the included services can be discovered on the target node or a neighbour within certain hop distance, through a process called service-binding. However, in both cases, the code of all tasklets or services must be part of the pre-installed node firmware. POBICOS [28] allows application-specific agents to be written in C, and supports a fully dynamic placement and migration of such agents in the WSN.

A recent trend are service-oriented architectures based on web services for which support is readily available in most popular programming languages. In the approach presented in [2], nodes publish their basic functionalities in the form of services to a base station, which in turn provides the necessary abstractions for accessing/manipulating these functions via a high-level programming language of the user's choice. TinyWS [26] allows direct HTTP-based access of node services through a SOAP processing engine, while the CoAP protocol [19] follows a RESTful design. These frameworks are easy to use, but do not support any intra-network processing.

Combining elements of the above approaches seems a more promising strategy. In [6] authors integrate TENET in a WSN-SOA to support task deployment and collaboration between nodes as a service. In USEME [4], applications can be composed using a declarative language which specifies nodes, groups and services; however, service implementation depends on the underlying OS. ProFun TG [12] is based on a declarative network programming model that utilizes abstract task graphs and data flows. A visual IDE is provided for handling tasks, but the user is also given the option of programming application-specific tasks in C or SEAL [11]. Task allocation and data handling is feasible through JSON requests over HTTP.

While the task-based approaches in [6, 12] and the multi-layered software architecture in [14] offer advanced programming features, user-defined software must be part of the initial firmware. In contrast, our work focuses on providing easy to use interfaces that allow flexible data manipulation, node reconfiguration and task reprogramming, even after deployment. An initial presentation of our approach can be found in [15]. This paper includes an updated and thorough description of our implementation, as well as an evaluation of the key client functions supported.

3 System Architecture

The high-level architecture of our approach is illustrated in Fig. 1, highlighting the most important elements of the system. The WSN consists of one or more sensor nodes and a gateway. The gateway acts as the front-end of the WSN towards external clients, offering an interface through which they can send configuration commands and queries, and receive data from the WSN. In addition, through the same interface, clients can deploy custom application tasks in the WSN and interact with them.

This functionality is implemented using two middleware-level components, running on the WSN gateway and on the sensor nodes, respectively, which communicate via a suitable middleware-level protocol. The gateway middleware translates client requests into corresponding messages that are sent to the sensor nodes. Conversely, it

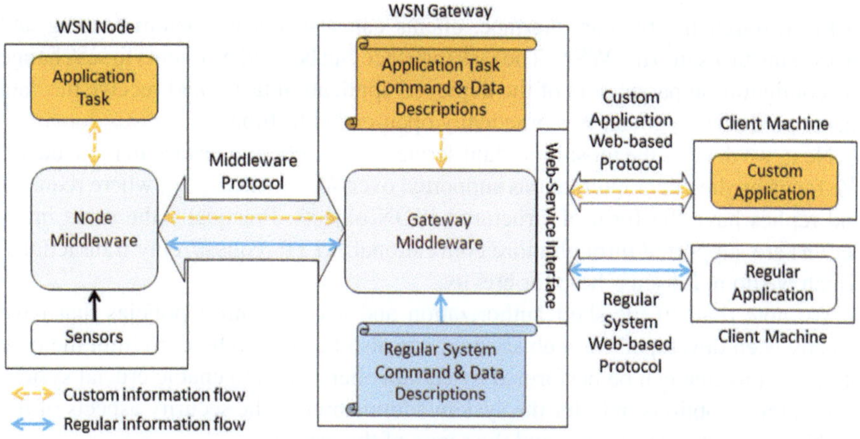

Fig. 1 High-level system architecture, for the simultaneous and uniform support of regular command/data flows and custom application command/data flows

translates the messages received from the sensor nodes into replies that are forwarded to the respective clients. The node middleware receives the messages coming from the gateway, handles the configuration commands, and prepares/sends messages with the corresponding replies and/or data produced. It is also responsible for installing and running custom application tasks on the sensor node, as well as for supporting their interaction with the external application clients, via the gateway.

As indicated in Fig. 1, on the WSN gateway, the translation between the client-level protocol and the middleware-level protocol is performed based on suitable metadata descriptions. This holds for the regular client functions of the WSN (which are fixed and a priori known) as well as for the commands and data of the custom application tasks that have been deployed in the WSN (which are open and can be defined in a dynamic way). The former metadata descriptions are part of the gateway firmware, whereas the latter are registered with the gateway when installing the application task. This way, the middleware can use the same protocol translation engine and middleware protocol for both information flows.

4 Client Interface

The interface provided by the gateway to external clients follows a web-based design. Given the wide support of web services in all popular programming languages, this allows the WSN to be accessed practically from any platform in a convenient way. The client interface consists of two parts: the regular system interface, and the custom task interface. The regular interface allows clients to configure the sensor nodes, issue queries and receive sensor measurements (without using any custom application task). Through the regular interface, clients can also install custom sensing and processing tasks into the WSN. The task interface can be used by clients to set/change the configuration parameters of the deployed application tasks, and receive the data that is produced by them, in a practically open-ended fashion.

Next, we discuss the most important services of the client interface in more detail. We focus on the client interactions supported over WebSockets [29], where requests and replies have the form of structured JSON objects. Practically the same interactions are supported through more conventional HTTP request/reply transactions, which we do not discuss here for brevity.

We note that off-the-shelf authorization and access control policies that have already been developed for web services can also be applied here, to limit the type of operations that can be performed by certain clients, and to enable crucial system configuration options only for the system administrator. The security aspects of the WSN access interface are beyond the scope of this paper.

4.1 Node Status and Configuration

Monitoring and fine-tuning key operational parameters of the WSN can be crucial in achieving robust and efficient system operation. To confirm their liveness, sensor nodes periodically send status messages to the gateway. These messages contain various information about the current state of the node, such as the transmission power, the battery voltage, the quality of the wireless link with the gateway, the application tasks installed, and the elapsed time since the last node reboot/reset. Status messages are produced as part of the default operation of the WSN, without any explicit request from any client.

Node status information is collected and stored at the gateway, in order to have an updated view of the WSN and to track its operation over larger periods of time. This information can be retrieved by external clients via a corresponding request, in the spirit of the first row of Table 1, where the client requests the most recent status information for the node with address 12.25, and receives certain status information as a reply.

Clients can also set various configuration parameters of the sensor nodes, such as the transmission power or the status reporting period. As an example, the second row of Table 1 shows a configuration request for node 12.25, which sets the status reporting period to 60 s and the transmission power at level 28. Configuration requests are explicitly acknowledged via a reply message. In case the requested setting was not successful (e.g., if it is not supported by the node or the node does not respond), an error message is returned with a brief description of the failure. Note that clients should have at most one pending/unacknowledged configuration request at any time, i.e., a client should not send the next configuration request before receiving a reply for the previous one.

Table 1 Node status and configuration

Node status	{addr:[12,25],service:'node_stat'}
Reply/data	{addr:[12,25], timestamp:754, firmware:1, statperiod:60, txpower:31, battery:3.05, rssi:60, lqi:107, deluge:off, time:2015:09:18:23:36:58}
Node configuration	{addr:[12,25],service:'node_config', set:{statperiod:60,txpower:28}}
Reply/data	{addr:[12,25], node_config:OK}

4.2 *Sensor Queries*

Clients can query the WSN to receive sensor measurements in two different ways. The first option is to issue a one-shot query. An example is given in the first row of Table 2, for a query targeting the temperature and humidity sensors of the node with address 12.25. In this case, a single reply will be produced with the requested data, as shown in the third row of Table 2. The reply also contains additional node status information, in the spirit of what is already shown is Table 1, not repeated here for brevity. If the target node does not have the specified sensors, an error message will be returned instead. Notably, the entire WSN (all sensor nodes) can be addressed by using a special broadcast address in the query request, in which case several distinct data objects will be produced as a reply, one for each node.

The second option is to submit a long-running query in order to perform periodic sensor measurements. The second row of Table 2 gives an example for obtaining measurements from the temperature and humidity sensors of node 12.25, with a sampling rate of 60 s. As a result, multiple replies will be produced and sent back to the client over the WebSocket, each having the same form as the reply for the one-off query (see above). Periodic queries remain active as long as the respective WebSocket connection is kept open. When the client closes the WebSocket, the gateway automatically cancels all queries that were issued through it, and releases the respective system resources in the WSN.

Table 2 Sensor queries and data produced

One-shot query	{addr:[12,25], service:'query', mode:'oneshot', sensors:{temp, hum}}
Periodic query	{addr:[12,25], service:'query', mode: 'periodic', sensors:{temp:60, hum:60}}
Reply/data	{addr:[12,25], ..., values:{temp:24.35,hum:56.267}}

4.3 Installation of Custom Application Tasks

The gateway offers a set of services so that clients can deploy custom application tasks in the WSN in a dynamic way. The first step is to upload to the WSN gateway the code of the task, along with the task-specific descriptions for the configuration commands that can be handled by the task and the data objects that will be produced by it. As will be described next, these descriptions are XML files that must follow a specific syntax. The file upload can be done by using standard HTTP/FTP support.

Before task installation actually begins, the installation service has to be activated on the target sensor node. Then, the client can initiate the installation of the application task. When this completes successfully, the task will automatically start running on the sensor node. Finally, the installation service needs to be deactivated. Table 3 shows indicative requests for these three steps.

Note that the installation service has to be activated and deactivated in an explicit way. This is to avoid having the corresponding program running continuously in the background on the sensor node—even when idle, this still takes up memory, and might also incur some processing and traffic in the WSN. If this optimization is not important for a given deployment, the sensor nodes can be configured to keep the task installation service active at all times, and the start/stop requests can be skipped or completely removed from the WSN client interface.

4.4 Task Configuration and Data Retrieval

Once a custom application task is installed on one or more sensor nodes, the client can interact with it, again using a web-based request/reply interface. Note however that each task can have its own application-specific configuration parameters and produce its own application-specific types of data. Since the contents of the respective service requests/replies are not known in advance, they are defined by the application developer, via suitable XML descriptions that follow the rules of the DTD's shown in Table 4.

Table 3 Custom application task installation

Start installation service	`{addr:[12,25], service:'task_install', action:'start'}`
Install task	`{addr:[12,25], service:'task_install', action:'install', file='myapptask'}`
Stop installation service	`{addr:[12,25], service:'task_install', action:'stop'}`

Table 4 DTDs for task-specific command and data object descriptions

Task commands	<!ELEMENT commands (command*)> <!ELEMENT command (uname,code,par*)> <!ELEMENT uname (#PCDATA)> <!ELEMENT code (#PCDATA)> <!ELEMENT par (uname,valtype)> <!ELEMENT valtype (#PCDATA)>
Task data objects	<!ELEMENT datamessages (datamsg*)> <!ELEMENT datamsg (uname,code,data*)> <!ELEMENT uname (#PCDATA)> <!ELEMENT code (#PCDATA)> <!ELEMENT data (uname,valtype,size?)> <!ELEMENT valtype (#PCDATA)>

Task configuration requests take as parameters the identifier of the command, followed by zero or more parameters. The XML description should specify the name of the command identifier, the number and type of its parameters, and the order in which these should appear in the client request. The gateway uses this description to parse client requests and verify that they conform to the specification, before mapping them to the packets that will be sent to the sensor nodes. The order in which parameters are marshalled into the packet is the same as the one specified in the XML description.

Along the same lines, the data produced by the application task is defined in a flexible and structured way, via a corresponding XML file. This may contain one or more data messages, representing distinct data objects which in turn can include several data elements. Optionally, a size element can be used to indicate an array of values of the same type and name. The order in which elements appear inside the data object also defines the order in which the application task is expected to marshall the respective values when sending them to the gateway. At the gateway, the same order is used to unmarshall the values from the packet and to pick the proper name-tags that will be added to the data values when constructing the message that will be ultimately returned to the client.

Similarly to periodic queries, the data objects that are produced by the application task can be retrieved through a WebSocket which the client keeps open with the WSN gateway. In addition, task data is stored persistently at the WSN gateway (even if a client retrieves it via a WebSocket). It can be downloaded, in a file, via a conventional HTTP/FTP request, at any point in time. If desired, the client can request for this data to be deleted upon a successful download.

We note that analogous persistence could be supported for regular periodic queries as well, to keep them active even if the client closes its WebSocket connection to the WSN gateway. However, clients would then have to cancel such queries in an explicit way (via a suitable request). This in turn could lead to orphan queries, and the need for a suitable garbage collection mechanism. No such issue exists for application tasks, which are deployed in the WSN for very long periods, and thus have a more permanent nature compared to ad-hoc regular client queries.

It is important to stress that the requests/replies and data objects of the regular client interface (see previous subsections) are also specified using such XML descriptions. Thus the same parsing and protocol conversion engine can be employed for both the regular and the custom task interface of the WSN. In fact, the regular interface can be changed/extended merely by replacing the respective XML descriptions, without modifying the WSN gateway software; but, of course, the middleware on the sensor nodes would have to be updated accordingly.

5 Task Programming Model

Custom application tasks are implemented as standalone software components, written in the C language. Tasks are purely event-driven, and do not have a control thread of their own. More specifically, the programmer must provide three event handlers, shown in Table 5. The *on_init*() handler is called when task execution starts, in order to perform initialization actions. The *on_sensordata*() handler is called when new sensor measurements become available, so that the task can process them as desired. The sensors for which values are available and the corresponding measurements are passed as arguments to the handler routine. Finally, *on_command*() is called when the application client issues a configuration command for the task. It takes as arguments the identifier of the command and a buffer with the marshalled values for its parameters in the order that is specified in the corresponding XML description.

Within these event handlers, the programmer can use a set of special primitives, also listed in Table 5. The *setSensorPoll*() primitive is used to specify the sensors to be polled on behalf of the task, and the sampling rate for each one of them. This is typically called within the *on_init*() handler so that the task will start receiving sensor values. In turn, the runtime system of the sensor node performs the desired sensor access, and up-calls the *on_sensordata*() handler to forward the values obtained to the task. Within that handler, *getSensorDataPos*() can be used to check the availability of a specific sensor value and retrieve its position in the value array that is passed as a parameter.

The task must store any data it wishes to send to the external application client in a byte buffer. The buffer has to be managed by the task according to its needs; it can be a global variable that persists across handler invocations, or a temporary variable used in the context of a single handler invocation. The type of the message buffer is set via *setDataMsgType*() to one of the data object identifiers defined in the XML description, and the payload is marshalled into the buffer using the *write* $*$ () primitives. The programmer is responsible for marshalling data in the order specified in the XML description, so that data can be properly unmarshalled at the gateway. Finally,

Table 5 The application programming interface for custom tasks

Event handlers	
void on_init(void)	Initialize task state
*void on_sensordata(int mask, float *vals[])*	Handle new sensor values
*void on_configcmd(char id, char *buf)*	Handle configuration command
Sensor primitives	
setSensorPoll(int mask, int rates[]) *getSensorDataPos(int sensorid, int mask)*	Specify the sensors to poll, and the sampling rates Use within *on_sensordata()* to check *mask* for the availability and the position of sensor data in *vals*
Outgoing data primitives	
*setDataMsgType(char *buf, int type)*	Set message buffer type
*writeUInt8(char *buf, int *pos, int v)* *writeUInt16(char *buf, int *pos, int v)* … *writeFloat(char *buf, int*pos, float f)*	Data marshalling routines
*sendDataMsg(char *buf, int len)*	Send data message to the gateway
Incoming command primitives	
*readUInt8(char *buf, int *pos, char *v)* *readUInt16(char *buf, int *pos, int *v)* … *readFloat(char *buf, int*pos, float *f)*	Data unmarshalling routines

when the task wishes to sent the prepared data message, it calls the *sendDataMsg*() primitive.

A similar approach, in the reverse direction, is followed for incoming configuration requests, which are communicated to the task via *on_configcmd*(). In this case, the configuration command identifier is automatically extracted and provided as a parameter, but the task remains responsible for unmarshalling the parameters from the message buffer, via the $read*()$ primitives. Again, this must be done in the order given in the corresponding XML description.

In order to reduce the amount of information that is sent over the wireless links, the only "type" information present in the configuration command and data messages, is the command and data message identifier, respectively. The marshalling primitives do not add extra information for the type of data that is written into the message buffer, and the unmarshalling primitives do not try to verify the actual type of the data being read out of a message buffer. As a consequence, no error will be returned if the application developer does not marshal/unmarshal data as per the XML descriptions. If desired, one may adopt a more type-safe approach, at the cost of adding extra information in the messages sent/received by application tasks.

6 Implementation

This section describes the most important elements of our proof-of-concept implementation. We focus on the middleware that runs on the sensor nodes and the gateway of the WSN. The wireless communication between the sensor nodes and the gateway is performed on top of IEEE 802.15.4, using our own middleware-internal protocol. The prototype currently supports a star network topology, where every sensor node is in reach of the gateway, but it is rather straightforward to extend the implementation to support multi-hop topologies by adding suitable ad-hoc routing functionality at the lower network layers of the system.

6.1 Hardware Platform

As a hardware platform for the sensor nodes we use TelosB nodes from AdvanticSYS (Fig. 2a), with 48 K of ROM and 10 K of RAM. For the WSN gateway, we use a mini-PC (Fig. 2b) connected over serial to a sensor node which acts as the root of the rest of the wireless sensor network. The mini-PC can be accessed from the Internet over a wireless or wired connection, depending on the installation.

6.2 Node Middleware

We have implemented the middleware that runs on the sensor nodes using Contiki [9], a popular open-source operating system for WSN applications, which has been ported to a wide range of hardware platforms. Contiki comes with strong network protocol support such as ContikiMAC [8] and Rime [7], offering different transport options. It also has ready-to-use software for over-the-air file transfers via the Deluge protocol [16], and for persistent file/data storage on Flash memory using the Coffee file system [27]. Last but not least, it is possible to dynamically load and run ELF-based code on RAM, without having to reprogram the whole firmware or to reboot the node.

Fig. 2 Hardware used to build the system prototype

(a) Wireless sensor node (b) Gateway mini-computer

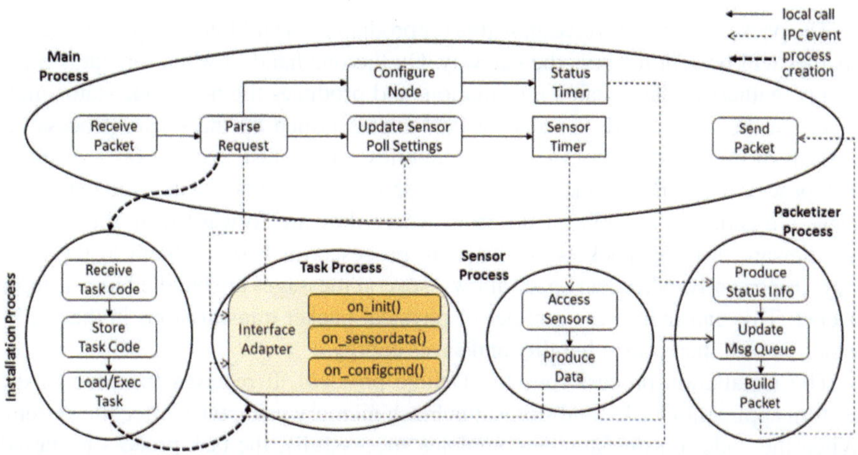

Fig. 3 Architecture of the sensor node middleware. The code of custom application tasks, which is dynamically installed on the sensor node, is highlighted

The node middleware implements the access services provided by the WSN gateway to external clients, on the sensor nodes. Following the Contiki programming model [10], the node middleware is structured as a set of distinct processes, as illustrated Fig. 3. These processes run concurrently and interact with each other via the Contiki OS: they wait for, handle and generate IPC messages, also referred to as events (indicated in the figure via the dashed lines).

The main process takes over the communication with the WSN gateway. When a middleware protocol packet is delivered from the Contiki network layer, it is parsed to extract and process the request. Node configuration commands are handled by the main process, which is also responsible for setting a timer that triggers the production of the node status information. Task configuration requests are forwarded to the task process. In case of a query request, the main process updates an internal table with the sensors that need to be sampled and the respective sampling frequencies, and sets a timer accordingly. When the timer fires, an event is sent to the sensor process, indicating the sensors that need to be sampled (this happens immediately in case of a one-shot query request). Finally, the main process starts (and stops) the installation process, on demand, when it receives the respective client requests.

The sensor process waits for an event from the main process, and accesses each of the specified sensors (a single event may concern several sensors). The obtained measurements are forwarded, via suitable events, to the packetizer process and/or the task process, depending on the recipient(s) specified in the event. Sensor access is implemented in a synchronous way, by issuing a request to each sensor and waiting for the measurement to be returned. This works well for both on-board and externally connected sensors, provided the response times are short (this is the case in our prototype); if needed, the sensor process can be extended to support asynchronous access of sensors with a large response delay.

The packetizer process is responsible for producing the middleware protocol packets that will be sent towards the gateway. On the one hand, it waits for status timer events, gathers/updates status information, and produces the next node status message. On the other hand, it waits for data events issued by the sensor process (as a result of regular one-shot or periodic queries) or the task process (as a result of task processing), and produces the respective messages. Node status and data messages are added to a FIFO queue, from where they are forwarded, in the form of middleware protocol packets, to the main process which sends them to the gateway. Importantly, whenever space allows, several messages are bundled into a single packet. This can reduce the number of physical packet transmissions in the WSN, which in turn increases reliability and saves energy.

The installation process runs the Deluge protocol. It receives the code of the custom application tasks, and stores it in Flash memory using the Coffee file system. When the code transfer/storage completes successfully, the task process is started in order to execute the application task—the dynamic loading of the task code from Flash to RAM is done by the Contiki ELF loader. If the task process is already running (executing a previously loaded task), it is stopped and then restarted again to load/execute the new application task.

Finally, the task process takes over the execution of the custom application task. The interaction between the task and the other processes of the node middleware is performed through suitable adapter logic, which is inserted in the task code by a preprocessor, before producing the final ELF binary. In a nutshell, after initialization and the invocation of the *on_init*() handler, the task process waits for configuration command events or sensor data events, and invokes the *on_configcmd*() and *on_sensordata*() handler, respectively (see Sect. 5). When the task asks for certain sensors to be polled via the *setSensorPoll*() primitive, a request event resembling a periodic query request is sent to the main process in order to update the sensor polling settings. Also, when the task calls the *sendDataMsg*() primitive, a corresponding event is issued to the packetizer process.

6.3 Gateway Middleware

The gateway middleware consists of two software components, running on the root node of the WSN and the mini-computer that acts as the front-end server of the WSN, respectively. The communication between these two components is over a serial port. To achieve a better separation of concerns but also to simplify debugging, this data exchange is kept independent of the middleware-level protocol used within the WSN, and is done in ASCII. More specifically, the requests sent by the server have the form of shell-like commands, and the replies/data sent by the root node are sequence of byte values in fixed-width hexadecimal format. The architecture of the gateway middleware is illustrated in Fig. 4, and is described in more detail below.

Like the "proper" sensor node middleware, the middleware of the root node also runs on top of Contiki. It is structured using three processes, as shown in the left

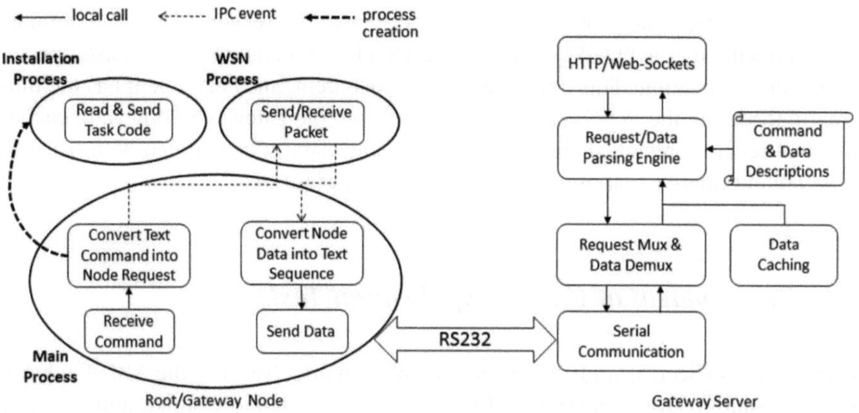

Fig. 4 Architecture of the WSN gateway middleware, on the root/gateway node and the gateway server

side of Fig. 4. The main process is responsible for the serial communication with the server, handling incoming commands as needed. If the command is addressed to one or more sensor nodes, it is converted into a request as per the middleware protocol, and is forwarded to the WSN process for transmission to the node(s) over the 802.15.4 radio. If the command concerns the root node, it is processed locally by the main process. The commands for the activation and deactivation of the task installation process are forwarded to the WSN process but also lead to respective actions locally. When the WSN process receives a packet from a sensor node, it is forwarded to the main process, which in turn sends the payload to the server in ASCII form. If a packet contains several piggy-backed messages, these are extracted and handled one by one, as if they had been received separately.

The server-side gateway middleware runs on top of a standard Linux environment, and is based on the Python CherryPy framework. Its high-level structure is illustrated in the right side of Fig. 4. One of the key functions is to convert the client requests, received over HTTP and WebSockets, into shell-like commands that are sent over serial to the root node, and, conversely, to transform the data received from the root node into HTTP/JSON replies for the respective clients. This data transformation/conversion is driven, in an open and fully parameterized manner, via the XML descriptions of the regular and custom task interface.

Notably, several clients can be connected to the gateway server to issue requests and queries. The gateway middleware keeps track of all outstanding client requests/queries. When a new request arrives, it checks whether this overlaps with requests that have already been propagated to the WSN. In case of a complete overlap, the new request does not lead to any action towards the WSN. Else, a new command is sent to the root node, for the parts of the request that are not yet covered. In the reverse direction, the replies/data received from the root node are fanned-out to the clients that have issued corresponding requests. As a result, the burden on the WSN is greatly reduced when several clients issue the same configuration/query requests—in effect, only one request will be sent to the sensor nodes, and the data produced will be distributed to all the interested clients, in the spirit of a subscription-based communication. This optimization is transparent, and each client has the illusion of accessing the WSN on its own. In addition, the gateway middleware caches the most recently received status information and data in case it can be re-used to satisfy client queries that might be issued in the near future.

6.4 Development of Custom Application Tasks

As discussed in Sect. 5, application tasks are written in C, based on the API of Table 5. It is important to stress that this API is not tightly coupled to Contiki, and could be supported on top another OS, like TinyOS [22]. Hence tasks remain independent of the underlying OS, and the application developer does not need to know how to write programs for the Contiki environment.

The code of the application task is passed through a pre-processor, which generates the source code of the Contiki program that will run on the node. This is basically the code that was provided by the application developer, plus some adapter code that implements the interface with the rest of the middleware (as was previously discussed). Using the off-the-shelf Contiki development tool chain, the code is compiled into a dynamically loadable ELF binary. Last, in order to deploy the task in the WSN, the ELF file, together with the descriptions for the task's commands and data, have to be uploaded to gateway.

7 Greenhouse Monitoring Application Example

At the Institute for Research and Technology Thessaly of CERTH, we are developing algorithms and systems for the efficient climate control of greenhouses. The objective is to optimize plant growth and development, as well as to reduce production costs for cooling/warning, water supply and fertilizers. This work is systematically evaluated using a testbed, shown in Fig. 5a, consisting of three greenhouses where various technological and monitoring and control approaches can be tested for different crop types in a flexible way. Each greenhouse is equipped with its own sensing and actuation infrastructure, which is connected to an external computer system that collects sensor measurements, processes this data, takes control decisions so as to maintain the desirable climate conditions and to schedule irrigation in a suitable way, and drives the actuators of the greenhouse accordingly.

Climate control in greenhouses conventionally relies on measurements taken from a sensor unit that is located at the center of the greenhouse. This approach assumes complete homogeneity of greenhouse microclimate (a perfectly stirred model). However, this assumption is not generally valid, and even less so for modern greenhouses with sizes that have greatly increased over the last decade. Recent research [3] has

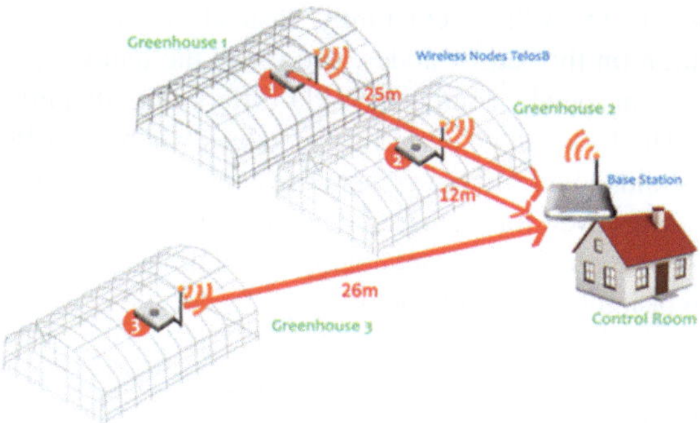

(a) High-level view of the testbed.

(b) Wireless sensor node placed near a plant.

Fig. 5 Greenhouse testbed of IRETETH/CERTH

in fact proven that greenhouse climate is extremely heterogeneous, with numerous exchanges occurring at the surface of the leaves, structure and equipment, largely governed by "local" climate, which varies also due to outside weather conditions and actions like opening or closing heating/ventilation systems.

For these reason it is important to adopt a more distributed approach, taking measurements at different locations in the greenhouse. This is not only technologically feasible, but it is also becoming increasingly attractive from a cost perspective due to the rapid advances of ICT technology. We are experimenting with WSN-based solutions, whereby several sensor nodes are deployed at various locations in the greenhouse, to measure different key environmental parameters. In fact, WSN nodes can be placed very close to the plants, as shown in Fig. 5b, to take measurements at an extremely localized scale. The information from the WSN feeds into a model that controls the greenhouse environment and irrigation in order to achieve optimal crop growth conditions with sustainable reduction of the irrigation water. Of course, such sensor nodes cannot be mains-powered, and have to operate on batteries. To increase their operational autonomy it it important to minimize the amount of data transmitted over the wireless network. This can be achieved, in a flexible way, using the system that was described in the previous sections.

As a concrete example, for tomato crops, we employ a model for scheduling irrigation, based on the Vapor Pressure Deficit (*VPD*) value inside a greenhouse [17]. The VPD is calculated as a function of temperature (*T*) and relative humidity (*RH*), based on the following formula:

$$VPD = 0.61 \times \exp\left(17.27 \times \frac{T}{T + 237}\right) \times \frac{RH}{100} \qquad (1)$$

The conventional way to drive this model is for the user/administrator to configure the sensor nodes to report the temperature and humidity values at a given period, and then calculate the VPD on the machine that runs the model. But this is quite inefficient, especially if significant changes in the VPD occur rarely. Instead, one can use a task that performs this calculation on the sensor nodes, and sends the temperature and humidity values only when the newly computed VPD differs (significantly) from the last calculated value. The task can also send these values when the VPD crosses certain application thresholds to inform the greenhouse controller as early as possible about such transitions. Moreover, for increased flexibility, the task can have a configurable notification threshold, which can be set/changed at runtime, according to the needs of the control loop or interactively by the user.

Listing 1 Code for the VPD task.

```
#define SETTHRESHOLD_CMD 1
#define UPDATE_MSG  1
#define DEFAULT_DIFF 0.5
#define TEMP_SENSOR 1
#define HUM_SENSOR 3

static float diff=DEFAULT_DIFF, tmp, hum, lastvpd;
static char mbuf[1+2*sizeof(float)];
static int cnt,report;

float calcVPD(float t, float rh){  }

on_init() {
    setType(mbuf, UPDATE_MSG);
    int rates[]={60, 60}; cnt=0; report=0
    setSensorPoll(TEMP_SENSOR | HUM_SENSOR, rates);
}

on_configcmd(int id, char *buf) {
    int pos=0;
    if (id == SETTHRESHOLD_CMD){
         readFloat(buf, &pos, &diff);
    }
}

on_sensordata(int mask, float vals[]){
    int pos, update=0; float vpd;

    pos = getSensorDataPos(TEMP_SENSOR, mask);
    if (pos > 0) { tmp = vals[pos]; update = 1; }
    pos = getSensorDataPos(HUM_SENSOR, mask);
    if (pos > 0) { hum = vals[pos]; update = 1; }
    if (update) {
        vpd = calcVPD(tmp,hum);
        if (abs(vdp - lastvpd) > diff) {
        report=1;
        lastvdp = vpd;
        else if(vpd < 0.5) { cnd=1; }
    else if((vpd > 2.0) && (vpd < 3.0)){ cnd=2; }
    else if((vpd > 3.0) && (vpd < 4.0)){ cnd=3; }
    else if((vpd > 4.0)){ cnd=4; }
```

```
        if ((prv_cnd != cnd) || report ){
            pos = 1;
            writeFloat(mbuf,&pos,tmp); pos += sizeof(
                float);
            writeFloat(mbuf,&pos,hum); pos = 0;
            sendData(mbuf, pos);
            report=0; prv_cnd=cnd;
        }
    }
}
```

Listing 2 XML description for the configuration command of the VPD task.

```
<command>
    <uname>setthreshold</uname>
    <code>1</code>
    <param>
        <uname>absdiff</uname>
        <valtype>float</valtype>
    </param>
</command>
```

Listing 3 XML description for the data object produced by the VPD task.

```
<datamsg>
    <uname>update</uname>
    <code>1</code>
    <data>
        <uname>tmp</uname>
        <valtype>float</valtype>
    </data>
    <data>
        <uname>hum</uname>
        <valtype>float</valtype>
    </data>
</datamsg>
```

Listing 1 gives the code for such the application task, based on the programming model that was described in Sect. 5. Listings 2 and 3 show the XML descriptions for the configuration command of this task and the data that is produced by it, respectively. The task code is 46 lines long, and compiles into a binary of roughly 2 K bytes, out of which about 1 K is the code of the API added by the pre-processor. Assuming the task runs on the sensor node with address 12.25, the client can set/change the notification threshold via a request `{addr:[12,25],service:'task_config', setthreshold:{absdiff:0.95}}`, and receive data in the form `{addr:[12,25], time:'2015:09:18:23:36:58', update:{tmp:31.01,hum:52.16}}`. One can imagine different versions of such a task with a more complex application logic and commands for setting additional configuration parameters, which can also produce more than one types of data objects. Note that using our system it is trivial to replace the task at any point in time, e.g., to correct a bug or to deploy a more advanced version.

8 Evaluation

We evaluate the performance of our implementation via a set of micro-benchmarks for typical primitive operations, summarized in Table 6. Note that the Config and Query operations follow a request-reply pattern, whereas Notify is a one-way communication from the sensor node to the client as a result of a long-running query. In the experiments, client requests are sent to the gateway in HTTP format, and the replies are returned using a JSON format (same as the one for the WebSocket interface). The measurements reported here are for a single sensor node that is accessed via the gateway. In the Query and Notify operations, the messages sent by the sensor node towards the client(s) include the temperature and the humidity values of the sensor node along with the node status information. The client is written in Python, and runs on an off-the-shelf laptop that communicates with the gateway over a fast Ethernet link.

Figure 6a shows the end-to-end latency of each operation, broken down to its main components: (i) the interaction between the client and the gateway, (ii) the interaction between the gateway and the sensor node, and (iii) local processing that takes place on the sensor node. As can be seen, Config and Query have the same client-gateway and gateway-node latencies. Also, the Query operation leads to significant processing

Table 6 Operations used in the performance measurements

Mnemonic	Description	Direction
Config	Set a node-level configuration parameter	Client ⟷ Node
Query	Request and retrieve a sensor value from the node	Client ⟷ Node
Notify	Receive a sensor value notification/update from the node	Client ⟵ Node

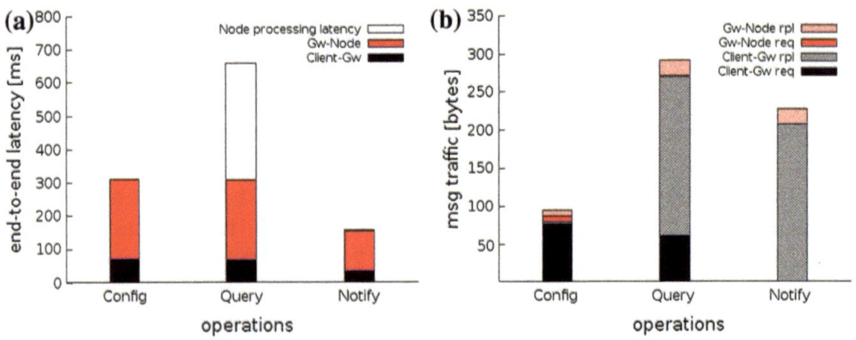

Fig. 6 Performance measurements for different primitive operations: **a** latency and **b** messaging cost of the respective interactions

at the node in order to schedule and perform the requested sensor measurement, whereas the Config operation is handled practically instantly. For the one-way Notify operation, the client-gateway and gateway-node latencies are about $1/2$ the ones observed for the request-reply operations. Unlike in the Query operation, in this case the node does not need to wait for the sensor(s) to be accessed, and immediately communicates the sensor values to the gateway when these become available, with a processing delay of less than a millisecond.

The total size of the messages exchanged in each case is shown in Fig. 6b. It can be clearly seen that the traffic between the gateway and the sensor node is much lower than the one between the gateway and the client. This is because the client-gateway interaction is done via HTTP/XML/JSON, whereas the gateway interacts with the sensor nodes using a compact binary message format. The Query operation is slightly heavier than Notify due to the extra request message, but the difference is small as in both cases the reply/notification message sent by the node (which contains the sensor values and the status data of the node) represent the bulk of the traffic. Note that the reply of Config is merely a simple acknowledgement, which is why this operation leads to significantly less traffic than the other two.

Further, we evaluate the overhead of query de-multiplexing on the gateway. We let a number of clients issue a long-lived query for temperature and humidity values with a period of one second, and record the time its takes for the gateway to forward to all clients the sensor value updates received from the sensor node. We also record the message traffic for the node-gateway and gateway-client interactions.

Figure 7a shows the de-multiplexing overhead in the form of a box-whisker plot, where the box boundaries indicate the minimum/maximum recorded values and the median is denoted by the line inside the boxes. The delay increases linearly to the number of clients, as expected, and remains small enough to be able to scale for more clients.

Figure 7b shows the number of messages produced in each scenario. Each notification involves a single message that is sent from the sensor node to the gateway,

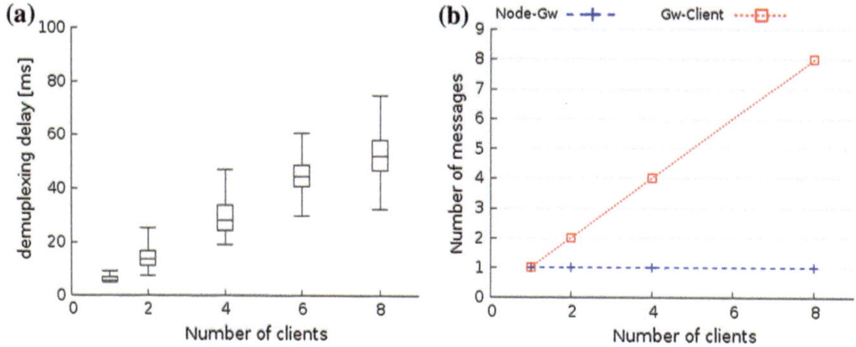

Fig. 7 Query de-multiplexing at the gateway for a varying number of connected clients: **a** client notification delay, and **b** node-gateway and gateway-client message traffic

which is then fanned-out to the different clients. Obviously, the latter traffic increases linearly to the number of clients. Note that without de-multiplexing, the same number of messages would also have to be sent from the sensor node to the gateway, introducing a significant communication burden in the WSN.

Finally, we evaluate the impact of performing computations locally on the sensor node, rather than sending individual measurements to the client for processing. As an indicative test, we take the example discussed in Sect. 7, where the input that is actually needed for the application to drive the control loop is the VPD. We compare two different modes of operation: (i) the "conventional" mode, where the sensor node periodically sends temperature and humidity values, and the VPD is calculated remotely, at the client; (ii) the "task-based" mode, where the client installs the custom VPD task on the sensor node, which performs the same measurements but only sends these values when the VPD difference exceeds a certain threshold or when the VPD crosses some application-specific values. In both cases, the notification messages that are sent from the sensor node to the gateway, carrying the temperature and humidity sensor measurements, are 9 bytes long.

Figure 8 plots the daily messaging costs for the two modes, based on measurements that were taken in one of our greenhouses over a period of 30 days. For the conventional mode, the sensor sampling and notification rate is 1 msg every 10 min (144 msgs/day), which is the default setting for our monitoring and control applications. Thus the messaging cost is constant, at almost 1300 bytes/day. To keep the comparison fair, the notification rate for the task-based mode is derived by post-processing (using the logic of the VPD task) the same temperature and humidity time series that were obtained in the conventional mode. It goes up to 20 msgs/day, at a cost of up to 180 bytes/day. But note that the average notification rate is much lower, and for some days it even drops to zero. Overall, the task-based mode leads to a substantial reduction of the communication between the sensor node and the gate-

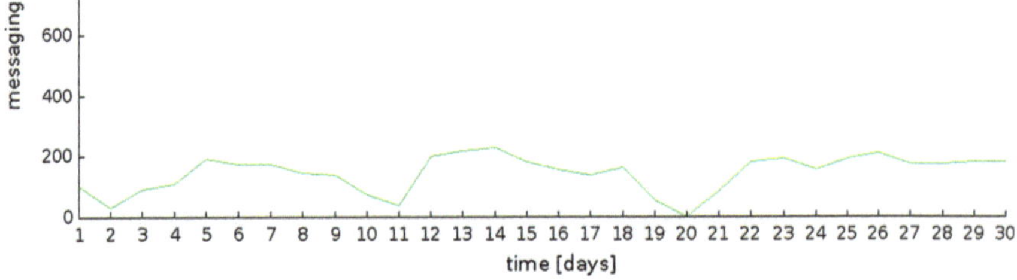

Fig. 8 Messaging cost of the conventional and task-based mode for a month

way, transmitting about 1/10 of the data sent in the conventional mode. The gains are so large that the overhead of deploying the task code on the sensor node can be amortized within a single day of operation.

9 Conclusion

We presented a unified service-oriented access to a WSN for regular queries as well as for the interaction with custom application tasks that are deployed in the WSN dynamically. This greatly increases flexibility, without introducing extra interfacing complexity for the clients of the WSN.

One of the enhancements we wish to pursue in the future is to extend the task programming model to support more advanced in-network processing, including data aggregation over multi-hop paths towards the gateway. We also wish to experiment with mechanisms for adapting the operation of the sensor node, in an application-aware manner, to further reduce the energy consumption of sensor nodes. Another direction is to investigate the scalability potential of the gateway for a very large number of clients, based on micro-server platforms.

Acknowledgements This work was funded in part by the General Secretariat for Research and Technology of Greece, action KRIPIS, project Smart Pole for Specialization and Development of Thessaly: Research, Innovation, Strategies.

References

1. Alcaraz, C., Najera, P., Lopez, J., Roman, R.: Wireless sensor networks and the internet of things: do we need a complete integration?. In: 1st International Workshop on the Security of the Internet of Things (2010)
2. Avilés-López, E., García-Macías, J.A.: TinySOA: a service-oriented architecture for wireless sensor networks. SOCA **3**(2), 99–108 (2009)
3. Bartzanas, T., Kacira, M., Zhu, H., Karmakar, S., Tamimi, E., Katsoulas, N., Lee, I.B., Kittas, C.: Computational fluid dynamics applications to improve crop production systems. Comput. Electron. Agric. **93**(C), 151–167 (2013)
4. Cañete, E., Chen, J., Díaz, M., Llopis, L., Rubio, B.: USEME: a service-oriented framework for wireless sensor and actor networks. In: 2008 Eighth International Workshop on Applications and Services in Wireless Networks (aswn 2008). IEEE (2008)
5. Christin, D., Reinhardt, A., Mogre, P.S., Steinmetz, R., et al.: Wireless sensor networks and the internet of things: selected challenges. In: Proceedings of the 8th GI/ITG KuVS Fachgespräch Drahtlose sensornetze, pp. 31–34 (2009)
6. Corre, B.L., Leguay, J., Lopez-Ramos, M., Gay, V., Conan, V.: Service oriented tasking system for WSN. In: 2010 Developments in E-systems Engineering. IEEE (2010)
7. Dunkels, A.: Rimea lightweight layered communication stack for sensor networks. In: Proceedings of the European Conference on Wireless Sensor Networks (EWSN), Poster/Demo session, Delft, The Netherlands. Citeseer (2007)
8. Dunkels, A.: The contikiMAC radio duty cycling protocol (2011)

9. Dunkels, A., Gronvall, B., Voigt, T.: Contiki - a lightweight and flexible operating system for tiny networked sensors. In: Proceedings of the 29th Annual IEEE International Conference on Local Computer Networks, LCN '04, pp. 455–462. IEEE Computer Society, Washington (2004)
10. Dunkels, A., Schmidt, O., Voigt, T., Ali, M.: Protothreads: simplifying event-driven programming of memory-constrained embedded systems. In: Proceedings of the 4th International Conference on Embedded Networked Sensor Systems, SenSys '06, pp. 29–42. ACM, New York (2006)
11. Elsts, A., Judvaitis, J., Selavo, L.: SEAL: a domain-specific language for novice wireless sensor network programmers. In: 2013 39th Euromicro Conference on Software Engineering and Advanced Applications. IEEE (2013)
12. Elsts, A., Bijarbooneh, F.H., Jacobsson, M., Sagonas, K.: ProFuN TG: a tool for programming and managing performance-aware sensor network applications. In: 2015 IEEE 40th Local Computer Networks Conference Workshops (LCN Workshops). IEEE (2015)
13. Fok, C.L., Roman, G.C., Lu, C.: Servilla: a flexible service provisioning middleware for heterogeneous sensor networks. Sci. Comput. Programm. **77**(6), 663–684 (2012)
14. Fortino, G., Guerrieri, A., Hare, G.O., Ruzzelli, A.: A flexible building management framework based on wireless sensor and actuator networks. J. Netw. Comput. Appl. **35**(6), 1934–1952 (2012)
15. Fronimos, T., Lalis, S., Koutsoubelias, M., Bartzanas, T.: Unified service-oriented access for WSNs and dynamically deployed application tasks. In: IEEE First International Conference on Internet-of-Things Design and Implementation, pp. 247–252. IEEE (2016)
16. Hui, J.W., Culler, D.: The dynamic behavior of a data dissemination protocol for network programming at scale. In: Proceedings of the 2nd International Conference on Embedded Networked Sensor Systems, SenSys '04, pp. 81–94. ACM, New York (2004)
17. Katsoulas, N., Kitta, E., Kittas, C., Tsirogiannis, I., Stamati, E., Sayvas, D.: Greenhouse cooling by a fog system: effects on microclimate and on production and quality of soilless pepper crop. Acta Hortic. **719**, 455–462 (2006)
18. Khan, I., Belqasmi, F., Glitho, R., Crespi, N., Morrow, M., Polakos, P.: Wireless sensor network virtualization: a survey. IEEE Commun. Surv. Tutor. **18**(1), 553–576 (2016)
19. Kovatsch, M., Duquennoy, S., Dunkels, A.: A low-power CoAP for contiki. In: 2011 IEEE Eighth International Conference on Mobile Ad-Hoc and Sensor Systems. IEEE (2011)
20. Leontiadis, I., Efstratiou, C., Mascolo, C., Crowcroft, J.: Senshare: transforming sensor networks into multi-application sensing infrastructures. In: Proceedings of the 9th European Conference on Wireless Sensor Networks, EWSN'12, pp. 65–81. Springer, Berlin (2012)
21. Levis, P., Culler, D.: MatÉ: a tiny virtual machine for sensor networks. SIGARCH Comput. Archit. News **30**(5), 85–95 (2002)
22. Levis, P., Madden, S., Polastre, J., Szewczyk, R., Woo, A., Gay, D., Hill, J., Welsh, M., Brewer, E., Culler, D.: Tinyos: an operating system for sensor networks. Ambient Intelligence. Springer, Berlin (2004)
23. Madden, S.R., Franklin, M.J., Hellerstein, J.M., Hong, W.: TinyDB: an acquisitional query processing system for sensor networks. ACM Trans. Database Syst. **30**(1), 122–173 (2005)
24. Noble, B.D., Satyanarayanan, M., Narayanan, D., Tilton, J.E., Flinn, J., Walker, K.R.: Agile application-aware adaptation for mobility. SIGOPS Oper. Syst. Rev. **31**(5), 276–287 (1997)
25. Paek, J., Greenstein, B., Gnawali, O., Jang, K.Y., Joki, A., Vieira, M., Hicks, J., Estrin, D., Govindan, R., Kohler, E.: The tenet architecture for tiered sensor networks. ACM Trans. Sens. Netw. **6**(4), 34:1–34:44 (2010)
26. Priyantha, N.B., Kansal, A., Goraczko, M., Zhao, F.: Tiny web services: design and implementation of interoperable and evolvable sensor networks. In: Proceedings of the 6th ACM Conference on Embedded Network Sensor Systems, SenSys '08, pp. 253–266. ACM, New York (2008)
27. Tsiftes, N., Dunkels, A., He, Z., Voigt, T.: Enabling large-scale storage in sensor networks with the coffee file system. In: Proceedings of the 2009 International Conference on Information Processing in Sensor Networks, IPSN '09, pp. 349–360. IEEE Computer Society, Washington (2009)

28. Tziritas, N., Georgakoudis, G., Lalis, S., Paczesny, T., Domaszewicz, J., Lampsas, P., Loukopoulos, T.: Middleware mechanisms for agent mobility in wireless sensor and actuator networks. In: Lecture Notes of the Institute for Computer Sciences, Social Informatics and Telecommunications Engineering, pp. 30–44. Springer Science Business Media (2012)
29. Wang, V., Salim, F., Moskovits, P.: The Definitive Guide to HTML5 WebSocket, 1st edn. Apress, Berkely (2013)
30. Yao, Y., Gehrke, J.: The cougar approach to in-network query processing in sensor networks. SIGMOD Rec. **31**(3), 9–18 (2002)

Towards Semantic Interoperability Between Internet of Things Platforms

Maria Ganzha, Marcin Paprzycki, Wiesław Pawłowski, Paweł Szmeja and Katarzyna Wasielewska

Abstract Lack of interoperability between Internet of Things (IoT) platforms, systems, and applications is one of serious problems that prevents its even more rapid adoption. As long as individual IoT platforms cannot meaningfully communicate with each other, the vision of a global IoT ecosystem will not be realized. While the problem of IoT interoperability can be considered at different levels, starting from hardware interoperability at the bottom of the stack, we are interested in the *semantic interoperability*, which can be seen as the meta-level for all interoperability considerations. In other words, the aim of this chapter is to consider how multiple IoT platforms can "understand" each other and have meaningful "conversation".

1 Introduction

One of the key factors, preventing full realization of the Internet of Things (IoT) [1] vision, is lack of interoperability between IoT platforms/systems/applications. As a matter of fact, today, many companies offer IoT systems to their customers, without perceiving any need to make them communicate with other such solutions. From the

M. Ganzha · M. Paprzycki · W. Pawłowski (✉) · P. Szmeja · K. Wasielewska
Systems Research Institute, Polish Academy of Sciences, Warsaw, Poland
e-mail: Wieslaw.Pawlowski@inf.ug.edu.pl

M. Paprzycki
e-mail: Marcin.Paprzycki@ibspan.waw.pl

P. Szmeja
e-mail: Pawel.Szmeja@ibspan.waw.pl

K. Wasielewska
e-mail: Katarzyna.Wasielewska@ibspan.waw.pl

M. Ganzha
Warsaw University of Technology, Warsaw, Poland
e-mail: Maria.Ganzha@ibspan.waw.pl

W. Pawłowski
Faculty of Mathematics, Physics, and Informatics, University of Gdańsk, Gdańsk, Poland

R. Gravina et al. (eds.), *Integration, Interconnection, and Interoperability of IoT Systems*, Internet of Things, DOI 10.1007/978-3-319-61300-0_6

commercial perspective, expenditure of resources needed to support interoperability, outside of the existing solution, is often perceived as unreasonable. This perception is particularly strong among SMEs (small and medium-sized enterprises) that we have contacted in the early months of 2016.

To counteract this process, of generating *IoT silos*, the EU has funded seven research projects that started in January 2016. Among them is the INTER-IoT project, which provides the background for the research reported in this chapter. The INTER-IoT project presents a comprehensive approach to IoT platform interoperability, offering its tools and services across the communication/software stack. Specifically:

- Device layer—seamless inclusion of new devices into the existing ecosystem of IoT devices,
- Network layer—handling of object mobility and information routing,
- Middleware layer—seamless service discovery and management of smart objects,
- Application service layer—reuse of heterogeneous services from heterogeneous platforms,
- Data and semantics layer—common understanding of data and information.

Let us observe that while our focus is on data and semantics interoperability, which is positioned on the highest level of the mentioned software stack, at the same time, semantics may be considered in regard to other layers. Specifically, the considered semantics necessarily needs to describe (at least provisionally) lower-layer entities, such as devices or services. Furthermore, solutions depend on lower layers to provide a full interoperability suite. In the envisioned architecture, the layers are independent when it comes to implementation, i.e. even though semantic methods might be used on other layers, the data semantics layer implementation does not depend on lower layers. In this work we focus solely on the highest layer and will not discuss in detail its interconnection to other layers.

The INTER-IoT project will provide methodology and tools that shall enable the building of the ecosystem of interoperable IoT platforms. Here, observe that while (for brevity) we will use the term *platform*, most of our considerations are applicable also to interoperability of IoT *systems* and *applications*. Therefore, unless it is explicitly stated, or follows from the context, these three terms are interchangeable. Let us now state some of the key top-level assumptions that guide our work.

1. While it is relatively easy to connect any two platforms, as this requires establishing a one-to-one translation, we are interested in multiple (three or more) platforms being combined to provide service for the user. This assumption removes a number of "easy path approaches", and makes the problem we deal with more realistic, and adds the challenge of scalability.
2. After a number of platforms is combined into a working ecosystem, it is possible that additional ones may need to be added. This process should not involve extra effort, larger than combining the original group of platforms.
3. Building the IoT ecosystem involves combining already existing solutions, which (likely) belong to different owners and have been developed using different,

including non-IoT, technologies (e.g. a Web Services-based application is combined with a graph database-based application which communicates using JSON messages and with an application that communicates using XML messages). Cooperation and communication between platforms should be made possible regardless of their underlying technology.
4. None of the stakeholders is ready/willing to introduce major changes to their applications to adapt to the needs of the others. Therefore, "integration" should be mostly done "outside" of their systems.
5. Our research is very pragmatic. We would like to use the existing artifacts (tools, ontologies, etc.) rather than creating something from the scratch. Furthermore, we recognize the fact that there are no IoT-related "accepted by majority" standards. Rather, the IoT space is littered with many small or large consortia, standardization bodies, and vendor unions. It is rather unlikely that any truly universal and global standards will be agreed to in the near future.

Based on these assumptions, here, we outline our proposal of an approach to reaching semantic interoperability [2]. To this effect we proceed as follows. We start from introducing the use case scenarios that underline the INTER-IoT project (in Sect. 2). Next, in Sect. 3 we summarize steps necessary to establish "common semantics" between multiple IoT platforms. This allows us to propose an architecture, which depicts our way of thinking about the IoT interoperability (Sects. 4 and 5). We summarize our approach in Sect. 6.

2 Use Case

To provide conceptual background of our work, let us briefly describe what are its real-world underpinnings. We assume that owners of multiple existing heterogeneous IoT "applications" decide to "combine them" to deliver a novel "service" to the users, or to enable exchange of meaningful information (to boost value provided by their applications). Upon reaching an agreement, concerning information that is to be shared, work commences to instantiate the needed service. We assume that platforms communicate through messages, therefore, the key is to make them understand information that they exchange (preferably, without introducing any changes to participating software artifacts). In other words, they should be sending and receiving messages in the data semantics and format that they understand. To achieve this goal, an additional software component that will be providing connection to the interoperability framework and the translation functionality, will be developed.

While the considerations presented here are general and apply to *any* group of IoT platforms, let us narrow down our interest further. The INTER-IoT project focuses on two application domains: *eHealth* and *Transportation and Logistics*. Achieving interoperability between platforms within both domains (and, also, between these domains) should be enabled by the results of the project. To better illustrate the needs and the expected results we will focus our attention on the *Transportation and Logistics* domain.

The considered location where multiple IoT platforms are to cooperate, is an international port, in which many companies (stakeholders) such as: container terminal companies, road haulier companies, and port management, operate. Each stakeholder is running, one or more, software systems (including IoT platforms) that collect data and, if this was planned during their deployment, are integrated with other systems. Note that, in the port itself, there are also: an IoT platform that controls lighting, and an IoT platform for management of data from meteorological sensors, pollution sensors, marine buoys, etc.

Lets now further focus on a specific scenario, for planning and monitoring transport services through interoperability of IoT platforms. Here, a road haulier company, which uses a fleet management software, provides a container transport services. The company receives information (orders) from the port management system, and it's system organizes truck routes accordingly (event-driven cooperation). Moreover, in planning activities, the haulier company uses information about the status of individual trucks, e.g. the location, the available driving hours, and the estimated time to finish the work already assigned. The haulier company has a part of it's fleet equipped with sensors provided by some software company, but also wants to include devices from other manufacturers that should interoperate with it's software. Separately, in its business practices, the haulier company (occasionally) subcontracts trucks from other truck companies and wants to (temporarily) consider them as part of their fleet. Moreover, its drivers can interact with devices in the truck and see selected pertinent data in an application on a mobile phone/tablet.

In this context, the overall aim of interoperability is to be able to exchange information about trucks, such as position, available driving hours, status. Here, trucks can be represented by virtual truck objects (cooperating with devices in physical trucks) available in the IoT platform cloud. Truck owners can provide access to subcontracted virtual trucks, so that they can receive orders from the road haulier company. The collected information should be available to: the road haulier company (communication of orders to the trucks), truck owner company (all data about the trucks), the driver (receive order and monitor his routes), the port management system (estimated time of arrival and attendance) and the container terminal platform (estimated time of arrival). Note that the range of information that is (can be) accessible depends on the interacting platform. The truck company platform has full information about its trucks, while the container terminal platform needs only trucks' estimated time of arrival. Interoperability should enable sending work orders directly to the truck company platform, from the haulier company platform. Obviously, we have to assume that it is extremely likely that each platform stores data in a different format and is using different semantics. Furthermore, it communicates, with third party systems, using different interfaces. For instance, let us assume that in one system data is stored in relational database according to the specified RDB schema and is exchanged with other systems via XML web services (with message format defined in an XSD), while in another platform data is stored in a NoSQL database and exchanged with other systems via a REST method with a JSON message format.

3 Data and Ontologies

Let us now discuss what needs to be done "on the way" toward semantic interoperability. First, let us introduce key terminology and underlying assumptions.

An *ontology* [3], in a broad sense, is a way of storing knowledge. It can contain information about both concrete data items about entities (individuals) and structural information about data, usually in a given area of interest. In the scope of this chapter, systems can be said to be semantically interoperable if any exchanged message can be (in a practical, not theoretical, way) expressed in terms of the ontology of the receiving system.

One formal way of expressing semantics is the OWL, our language of choice. It is a, W3C proposed, semantic web language designed to represent a complex knowledge from a domain by defining things and relations between them. It is used to represent ontologies, that are both human-understandable and machine-readable. Note that the OWL language may have different profiles [4], which represent different dialects of the *description logic* [5]. These profiles "trade" some expressiveness for the efficiency of reasoning. They include: OWL 2 DL, OWL 2 EL, OWL 2 QL, and OWL 2 RL. Additionally, there are different syntaxes [6] for persisting OWL ontologies including: RDF/XML, OWL/XML, Manchester, Turtle, and Functional. OWL ontologies are comprised of axioms—atomic pieces of information that contain formally defined knowledge. These axioms are mutually interconnected and state facts about a common set of entities.

OWL ontologies are often modular. There are two kinds of modularity. *Horizontal* modules are subsets of ontologies, often very loosely (or not at all) related that describe (very) different parts of an area of interest. A popular module in many ontologies is a geolocation module. In the scope of transport and logistics in a port environment, for instance, such module is independent of the structure of information about cargo containers. Even though both modules can (and likely will) be used to describe any given instance of a physical container, there is no formal connection between the location of a container and its type, weight, color, or any other property. Moreover, a geolocation module is not tied to this particular application, and may be used for many other physical entities, and even in entirely different domains. In summary, horizontal modules are not dependent on each other, and sometimes completely orthogonal.

Vertical modules, on the other hand, build on top of each other and form a hierarchy. Such hierarchy is a top-down structure, i.e. new modules are added "below" the old ones. This visualization corresponds to the level of abstraction encountered in hierarchies of ontologies, where most general ontologies are on top, and most specific ones on the bottom. A vertical module structure mimics this exactly. In fact, the difference between a vertical structure of modules, and vertical ontology hierarchy is largely terminological. If we adopt a perspective of treating ontologies as sets of axioms, and modules as subsets of axioms, there is, practically, no difference. In any case, vertical modularity, based on levels of abstraction, is a backbone of semantic interoperability. High-level ("top") ontologies (or modules) contain general terms that are "specified" by lower ontologies. An example that is central to the IoT, are "device" ontologies that describe, in general terms, IoT devices (sensors, actuators, smart and mobile devices etc.). These ontologies are then extended by domain-specific ontologies (often called just "domain" ontologies). In the eHealth domain, lower ontologies may add types of devices specific to healthcare, such as patient monitoring tools, automated medicine applicators and others. Transportation devices may include GPS sensors, truck speed sensors, port crane actuators and others. A canonical model of semantic interoperability assumes that the same high-level device ontologies are used in both examples. This means that, without additional effort, there is a set of general terms understood across domains.

For a much more detailed explanation of ontology modularity see [7].

3.1 Operations on Ontologies in the Context of Establishing Semantic Interoperability

Let us now introduce some key terms that are related to semantic interoperability. In the scope of this Chapter, we use the following definitions (cf. [8]).

An *ontology alignment* is a set of correspondences between two or more ontologies. Correspondences may be simple (between atomic entities) or complex (between groups of entities and sub-structures), but always relate entities from different ontologies. Alignments contain predicates about similarity, called a *matching* (e.g. equivalence and subsumption axioms), or a logical axiom—a *mapping*. Ontology alignment tools often state a degree of confidence for every correspondence in the mapping. Note, that the terms "mapping" and "matching" are often not distinguished in the terminology used by the alignment tools and are used interchangeably.

Ontology merging is a process of combining two, or more, ontologies into one so that the resulting ontology stores knowledge from all merged ones. A simplistic case of a merge is an ontology that is a result of a simple sum of sets of axioms from combined ontologies. More sophisticated merges include additional axioms that state how the axioms from combined ontologies relate to each other. Such additional axioms are often the result of an alignment.

Finally, *ontology (semantic) translation* is a process of changing the underlying semantics of a piece of knowledge. Given some information described semantically, in terms of a source ontology, it is transformed into information described in terms of a target ontology. Resulting information contains information interpretable (understandable) in the scope of target semantics. Semantic translation is considered good if the meaning of original knowledge is preserved. Ideally, no information is lost as a result of translation process. A perfect translation is also reversible, and original information can be ideally reconstructed by reversing the translation.

Our general approach relies on applying *ontology alignments* in the process of message translation (see Sect. 3.3). We assume that new instances of data may be generated dynamically by participating platforms. Therefore the structure (schema) that those instances comply to is of central interest to us. In other words, we aim to utilize alignments between structural information, rather than individuals. This enables a process of translation that is theoretically capable of translating any properly formatted message. The proposed message semantics translation is to work under the assumption that exchanged messages are ontologically demarcated and schema-compliant. The goal is to enable one-way or two-way "understanding" between software artifacts that implement differing semantics.

In order to consider a practical interoperability in a system consisting of multiple IoT platforms, we have to take into account a number of possible situations that stem from variable amount of semantic "readiness" in participating platforms. Here, the "simplest" case works under an assumption that every platform can provide an explicit *ontology* [3]. In this case, dedicated tools can be used to generate *alignments* between the participating ontologies [9, 10]. Working backwards from this scenario (towards "more difficult" situations), we consider cases where the the explicit ontology needs to be constructed/extracted (i.e., it is not yet available).

3.2 Extracting Ontologies

Besides ontologies, a given domain knowledge can be represented in other formalisms, such as relational database schema, XML schema, or JSON schema. They provide common understanding in terms of syntax rather than conceptualization, although some semantics can be deduced from the existing syntactic structure.

In order to provide common understanding of exchanged data, we decided to express semantics of platforms' that are to cooperate, as OWL ontologies persisted in RDF/XML syntax (officially recommended by the W3C). Therefore, we will either reuse existing OWL ontologies or transform semantics expressed in other formats to them. We have decided to consider, in the initial proposal, the following popular ways for IoT platforms to communicate data: (i) XML, (ii) JSON, (iii) RDF. In our research we have also considered the case when semantics has to be extracted from a relational database schema and expressed as an ontology. Let us now summarize our findings concerning tools for ontology extraction from schemas of these data formats. We will start from the most "friendly" and finish with the least friendly case. More extensive discussion can be found in [11].

3.2.1 RDF and RDFS

The Resource Description Framework (RDF[1]) is one of the W3C specifications for data interchange on the Web based on triples (subject, predicate, objects). The RDF structure forms a directed, labeled graph (RDF graph), where edges represent named

[1] https://www.w3.org/RDF/.

links between resources, represented by the graph nodes. The RDF Schema (RDFS[2]) is a way to describe structure of the RDF as it contains an extension to the RDF vocabulary, increasing expressiveness by, among others, introduction of taxonomies, domain and range for properties, etc. OWL ontologies are built on top of the RDF and the RDFS, adding more expressivity (semantics) that allows for more complicated reasoning. Ontologies in the RDF/XML syntax are directly compatible with the RDF, and can be processed by any RDF tool. In case of IoT platforms that exchanges data in the RDF format, the translation to OWL is natural, but not without caveats. Simple OWL ontologies can be directly "converted" into the RDFS, because OWL borrows a lot of core constructs from it; for instance: rdfs:range, rdfs:domain, rdf:type, rdfs:subClassOf, as well as multiple *xsd* datatypes. Other RDF or OWL constructs are more problematic, such as rdf:Seq (denoting a sequence) that does not have a direct correspondence in OWL. Nevertheless, mappings between RDF and OWL are well known [12]. This W3C recommended document describes the mapping of OWL 2 ontologies into RDF graph and vice versa (after satisfying some restrictions). When the RDF is serialized as RDF/XML and does not contain problematic constructs then it is already an OWL ontology. Conversion between ontology serialization formats can be done in tools such as: The University of Manchester OWL Syntax Converter[3] and Protege.[4] When problematic constructs appear, then the resulting OWL ontology is going to require extra attention of an ontology engineer that oversees the translation process.

3.2.2 XML and XML Schema

The XML[5] is the most widely accepted human- and machine-readable textual data format for exchange of information. The original method to define XML document structure is by using a Document Type Definition (DTD[6]). The DTD specifies the legal building blocks of an XML document as a list of elements and attributes, their ordering and nesting. Another method to model a document structure is by using the XML Schema (XSD[7]), which is written in XML, and provides more powerful means to define the structure and it's limitations (e.g. supports data-types and namespaces). It can be easily observed that both, the DTD and the XSD, can be mapped onto an OWL ontology specifying domain's conceptual model. Furthermore, XML instances can be mapped onto individuals built on that conceptual model. Mapping of the DTD to an OWL ontology was discussed, for instance, in [13, 14]. More work has been devoted to the design and implementation of mechanisms for transformation of XSD/XML into OWL (specifically into the RDF and the RDFS syntax). The popular

[2]https://www.w3.org/TR/rdf-schema/.

[3]http://mowl-power.cs.man.ac.uk:8080/converter/.

[4]http://protege.stanford.edu/.

[5]https://www.w3.org/XML/.

[6]https://www.w3.org/XML/1998/06/xmlspec-report.htm.

[7]https://www.w3.org/XML/Schema.

approach is to use XSLT stylesheets to store transformation rules, see [14, 15]. The case when the XML schema is not explicitly given can also be handled by some tools, by generating it automatically from one document [15] or from a corpus of documents [16]. Other approaches (discussed in [11]) were proposed in [17–19].

Besides the aforementioned research, there are tools for XSD and XML data transformation available online including: Ontomalizer,[8] ReDeFer.[9] Ontomalizer allows automatic conversion based on a set of predefined rules, even for very large schemas, e.g. the HL7 CDA (one of the test cases). ReDeFer is a set of utilities for different transformations e.g. XSD to OWL (based on XSLT transform) and XML to RDF, that can be tested in an online translation service.

Unfortunately, there is a problem with transforming "full" syntactic specification from XML to OWL. Specifically, constructs such as *sequence* and *choice* XSD elements cannot be directly translated to the OWL syntax. As a result, the information captured in ontologies does not always allow for the representation of OWL descriptions as XML documents valid with respect to a given schema. To deal with information-loosing translation of data from the XML to OWL and back, additional mechanisms to store such restrictions in an ontology have to be designed. This problem was addressed by authors of [20] where a framework for bidirectional transformations based on XS2OWL mapping model [21] is described.

Converting data, in the XML format, to the ontological representation is a crucial part of the methodology for preparing platforms for integration. We have identified a set of documented and partially available tools that can help in automating this task, however, at the end, the user verification is needed. It is necessary to inspect if all constructs have been properly converted due to the fact that there is an XML syntax that does not have an OWL equivalent, e.g. ordering of elements. Moreover, different tools handle different complexity of input XML/XSD documents, so the use of a given tool is dependent on how much expressivity is present in a given input document.

3.2.3 JSON and JSON Schema

JSON stands for "JavaScript Object Notation". It is a lightweight data interchange format, based on a subset of the ECMA-262 JavaScript standard object literal syntax. JSON has proven so practical that currently it is widely used for many purposes, well outside of pure web technologies. JavaScript objects are *dynamic* by nature, i.e. their structure can be dynamically modified in an essentially arbitrary way. In practice, however, developers are almost always interested in some ways of restricting this "mutability". The JSON Schema [22, 23] allows for describing the structure of the JSON data in a clear, human- and machine-readable way, enabling complete validation. Similarly to the XML Schema, it uses the same language as the data. An

[8]https://github.com/srdc/ontmalizer.

[9]http://rhizomik.net/html/redefer.

extension of the JSON Schema, which—among others—adds explicit treatment of hypertext links, is called JSON Hyper-Schema.

The JSON Schema website[10] lists many tools supporting the language, including validators, schema generators, and data-parsing libraries. JSON Schema documents can also be transformed into the XML Schema (XSD) format (e.g. using [24, 25]). Note, however that there is an intrinsic discrepancy between the two languages, which any such a conversion has to take into account. The XML Schema describes a "closed" content. Hence, instances of a schema can contain only items explicitly allowed by the schema. The JSON Schema, on the other hand, specifies an "open" content. Thus, schema instances—if not explicitly forbidden—may contain items beyond the ones requested by the schema. Therefore, while transforming the JSON Schema to the the XSD, one needs to make it open (e.g., by using the `<any>` element, wherever the JSON Schema does not prohibit additional properties).

In the context of our goals—lifting semantics to OWL—the following process is advocated. After transforming the JSON Schema document into the XML Schema format, any of the methods described in Sect. 3.2.2 can be used for obtaining an OWL ontology describing semantics of the (JSON) data.

If the JSON Schema is not available it has to be created or "discovered" from the data. Fortunately, there are tools available, which support this process. They can generate the schema based on single/multiple JSON document(s) and even additional "constraints" in the form of JSON schemas. More information about the tools and techniques for "discovering" semantics from "raw" JSON data can be found in [11].

3.2.4 RDB

Thus far we have deal with lifting, to OWL, semantics implicitly represented in messages. Let us now consider the same for stored data. Relational databases are a mature and still very popular technology. Compared to an ontology, a (relational) database schema is relatively "lightweight". It serves as a description of a structure of a specific database (in a formal language), i.e. how the data is stored; while an ontology represents structure of data in some domain, i.e. it describes a subject matter.

Database schemas are tailored for specific data storage solutions and are not usually meant to be reused in other software products. Schemas are usually described by a series of SQL statements, which can contain vendor-specific instructions (e.g. PostgreSQL, Oracle SQL, Microsoft SQL, etc.). Databases are optimized with respect to data storage and querying, while the actual "meaning" of data is inconsequential. This loss of meaning produces the core difficulty for capturing the semantics represented in the RDB.

Relational schemas can be represented by entity-relationship model diagrams (ER diagrams) that are usually constructed before a database implementation. A point of contention is that database relationships (i.e. database keys) are not named in the actual databases. Moreover, naming of relationships is optional in some ER diagram

[10] http://json-schema.org/.

standards (e.g. IDEF1X). This makes automatic creation of ontological models difficult. Traditionally databases do not contain any taxonomy—a stark contrast with ontologies, where taxonomies are the backbones.

Despite fundamental differences, a few methods of (semi)automatic construction of ontologies from database schemas have been conceived. Among the notable ones are DB2OWL [26] and RDBToONTO [27].[11] The latter algorithm includes sophisticated converters that attempt to infer information about the structure of the data (that is often implicit in RDBs) to build taxonomies.

In summary, the relational model may be too "sparse" for any automatic tool to recognize the structure and links between relational data. Nevertheless, while our experiences indicate that an ontology constructed by hand is always better than that constructed automatically from a database, the latter can serve as a good starting point for building a custom ontology. It can also be useful as a source for an alignment with other (e.g. popular and standardized) ontologies.

In summary, from the aforementioned research it can be concluded that there are tools/libraries that help transforming different data formats to ontologies. Nevertheless, in most cases, these tools allow only for the development of, what should be seen as, an initial OWL ontology. The inherent weakness of automatic ontology extraction, is that the resulting knowledge is self contained—it has no (automatically recognized) links to other ontologies. The produced ontology has then to be verified and, likely, corrected and improved by the ontology engineer. As a side observation, it is worthy to think about this fact when considering why the Semantic Web is not reaching maturity faster.

3.3 Finding Common Semantics

In this Chapter we discuss issues related to the interoperability of data exchanged between IoT platforms. Observe that two (or more) platforms can operate in the same domain, e.g. logistics, but use different semantics, e.g. one platform uses term *truck* with an attribute *capacity*, while the other uses the term *lorry* with an attribute *volume*. Obviously, these terms have the same meaning and can be treated as equivalent. This exemplifies the simplest case of an alignment—equivalence between two atomic terms. Observe also that capacity and volume may be represented in different units, e.g. one uses the metric system (cubic meters), while the other comes from US and uses cubic inches, which introduces a complication into the, otherwise simple, example.

As it should be obvious by now the in our considerations we do not deal with automatic discovery of meaning of incoming messages that were not agreed to "earlier". In other words, when multiple platforms are to work together, all messages that will be send/received have to be specified. With this specification in place, we can focus our attention on providing interoperability between semantics that is available (even

[11] https://sourceforge.net/projects/rdbtoonto/.

if only in an implicit format; see, Sect. 3.2) while forming the ecosystem. The way we propose to achieve interoperability is by facilitating translations (unidirectional or bidirectional) between cooperating platforms. This, in turn, will require creation of alignments between ontologies.

Let us now assume that for each platform that is to be joined, we were able to lift its semantics to the OWL (as specified in Sect. 3.2). The first step in building common understanding for such group of platforms is to understand how ontologies representing them "relate to each other".

While there are many ways, in which ontology aligning can be achieved, the simplest is: printing them, placing next to each other and establishing how concepts they capture relate. However, as soon as the ontology size increases, it becomes necessary to use available tools. Here, it should be stressed that potential size and complexity of ontlogies to be aligned leads us to question methods based on graphically representing and manually inspecting them. Spreadsheet-based alignment, such as FANPAI to SAREF mapping[12] may be a good starting point, but are not always viable and require an expert at every step. Instead, we looked for an answer to the question "how far can one go using automated tools", when it comes to ontology alignment for practical use.

In [9], we have summarized the state-of-the-art in the area of tools for ontology alignment/merging/translating, etc. Here, we outline the key findings. First of all, we have identified and inspected (from the point of view of their potential for *actual* use) 97 tools and frameworks that were described in research papers, or listed on websites devoted to ontology matching. The main conclusions of our work were rather discouraging. (1) Majority of these considerations are purely theoretical, without any available implementation. (2) Large number of tools stopped being maintained 1–2 years after they were completed. (3) Most of the significant tools were tested only during the annual OAEI [28] competition and the results were described in papers following it; however, many of top performers have been abandoned just after the competition. Overall, we have identified 9 tools that should be further evaluated as candidates to support semantic engineers in preparing alignments. The most interesting tools were LogMap [29],[13] COMA [30][14] and Agreement Maker [31].[15] For more details and description of other tools, see [9].

Furthermore, as a result of completed analysis, we have also decided to use the *Alignment Format* [32] as the base for alignment representation, in our proposed approach. To support our choice, let us also mention other methods for representing alignments between ontologies including: SKOS [33], MAFRA Semantic Bridge Ontology [34], OWL [35], C-OWL [36], SWRL [37], RIF [38], EDOAL [39].

SKOS (Simple Knowledge Organization System) is a vocabulary, in the RDFS, that allows to express relations between lightweight ontologies, e.g., taxonomies and

[12] https://docbox.etsi.org/workshop/2015/201504_smartappliances4/tno_study_smart_2013-0077_ldaniele.pdf.

[13] https://www.cs.ox.ac.uk/isg/tools/LogMap/.

[14] http://dbs.uni-leipzig.de/Research/coma.html.

[15] http://somer.fc.ul.pt/aml.php.

thesauri. All concepts in compared ontologies are identified by the skos:Concept class and described with properties such as broaderMatch, narrowerMatch, relatedMatch, closeMatch. This approach for expressing relations is intuitive, however, it is too simple for our scenario. Besides relations, we need to express also mapping rules with possible transformations (e.g. concatenation, split, unit conversion).

MAFRA Semantic Bridge Ontology (SBO) is an ontology, which instantiates a mapping specification, even though it is not an alignment data format. The main concepts, in the SBO, are *Service* (a function mapping arguments) and *SemanticBridge* (a relation between two sets of entities built from elementary services that are applied to them). Unfortunately, the complex format of the SBO ontology is described in the UML and tied to the MAFRA architecture.

OWL can also be considered as a language for describing relation between ontologies, e.g. by using equivalentClass, equivalentProperty, subClassOf and subPropertyOf. Drawbacks of this approach are: the correspondences (from the alignment), and declarations or definitions (from the original ontologies) are mixed, which makes the ontology more complex, and there is no way to express required data transformations (operation on data values).

Contextualized OWL (C-OWL) is an extension of OWL, for expressing mapping between ontologies. The new concepts are the *bridge rules* that describe directed correspondence between classes, properties and individuals. There are five relations that can be used: equivalent, disjoint, more general, more specific, overlap. However, the alignments in C-OWL are "too simple", since they are based only on named entities (no constructed classes can be used). Moreover, there is no way to express data transformations.

SWRL (Semantic Web Rule Language) is an extension of OWL with the concept of rule that can be interpreted as a correspondence between ontologies, when head and body parts come from different ontologies. Rule Exchange Format (RIF) is another format for specifying rules. The drawback of these approaches is that they are dedicated to rules and not alignments, e.g. they do not consider transformations.

The Alignment Format is a language independent format dedicated to persisting alignments in a simple and readable way. It was designed with the goal to provide common output format for matching tools. It is accompanied by an Aligment API [40], providing a programming library to persist and manage alignments expressed in the Alignment Format. The alignment document includes: declaration of ontologies, correspondences between entities from the ontologies with indicated type of relation and strength. The Alignment Format offers different levels, depending on the complexity of correspondence between entities (0—entities aligned by URIs, 1—replacement of entities pairs by sets of entities, 2—complex constructs handling with possible transformations).

The Expressive and Declarative Ontology Alignment Language (EDOAL) is an extension of the Alignment Format (level 2) that allows to specify rules for construction of target entities from source entities and various transformations that should be applied in the process (e.g. predefined operators, external functions call, XPath and regular expressions). EDOAL can be also manipulated with the Alignment API. Unfortunately, while the Alignment API provides functionality to parse and persist the alignment specification, it does not “execute” the logic encoded in these transformations. Therefore, even though the EDOAL is very expressive, all encoded transformation logic needs to be understood and implemented. According to our best knowledge, there is currently no tool that implements full support for EDOAL. Nevertheless, today Alignment Format seems to be the best solution for storing alignments. Additionally, the Alignment Format schema (DTD) is extensible, which leaves the door open for expansion of the range of information stored in an alignment, while still preserving a level of compatibility with existing Alignment API tools.

4 Towards Architecture for IoT Semantic Interoperability

Before proceeding further, let us summarize what has been discussed thus far.

1. We are interested in establishing semantic interoperability between multiple IoT platforms.
2. Since the “world of distributed applications” is still ways away from the Semantic Web, and since legacy applications are not based on ontologies and semantic data processing, we propose to use existing tools and approaches to lift semantics underlying such applications to the OWL. However, we acknowledge that all such ontology extraction has to have a “human in the loop”.
3. For the IoT systems that use different ontologies, after extraction/formalization, such ontologies can be aligned to build common understanding.
4. The Alignment Format (and related tools and formalisms) can provide a way of representing and storing relationships between ontologies underlying different IoT systems.
5. We intend to apply ontology alignments to facilitate the message translation process.

In what follows we use the term *translator* to describe a software artefact that is able to accept data expressed in one semantics and (with the help of alignments) produce data expressed in different semantics, but with the same meaning. The details of operation and architecture of a translator depend on a specifics of the complete solution. We discuss it in subsequent sections.

Let us now assume that a set of IoT systems that are to work together, has been “prepared”. For each one of them an explicit OWL ontology has been instantiated (or it already existed). How can we proceed to make them cooperate? Analyzing this situation we came to the conclusion that two main approaches can be proposed: (i) creating translators following each “flow of data exchange”, or (ii) use a common/central understanding and have a single translator for each IoT platform joining the ecosystem. The second approach can have different variants depending on the realization of the “central semantics”.

4.1 One-to-One Alignments

The first possible approach to connecting ontologies from heterogeneous IoT platforms is based on alignments between pairs of platforms. Specifically, alignments between platforms' semantics expressed as ontologies (original or obtained with methods mentioned in Sect. 3.2). In Fig. 1 a sample situation for four interconnected platforms (P, R, S, T) is depicted, however, it can be easily extended to any number of them. Prior to joining the interoperable ecosystem, an IoT platform administrator has to make a decision, which platforms she needs to communicate with, and whether the communication will be unidirectional or bidirectional. Here, note that in an actual application, it may not be necessary for all (sub)systems to talk to each other. Therefore, only the "needed" translators will have to be created. In the aforementioned logistics scenario we can imagine that the container terminal IoT platform is not going to communicate directly with the truck owner's platform (instead it communicates with the road haulier's platform), and driver's mobile application is only going to receive data from the road haulier platform (incoming work orders). Nevertheless, in Fig. 1, a "complete case" is presented—when all platforms need to communicate will all others. Each platform has a semantic data representation—ontology (OP, OR, OS, OT) that, if necessary, has been created prior to joining the ecosystem. Next, an

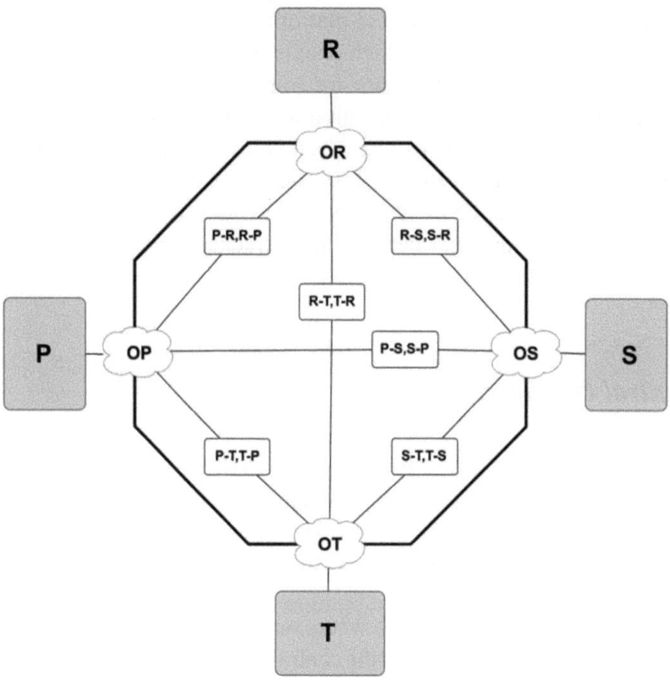

Fig. 1 One-to-one alignments

alignment is created between ontologies of each pair of platforms that are going to communicate. In case of bidirectional communication two alignments are need (a separate one to support communication in each direction), e.g. for P and R we have P-R and R-P.

Lets consider also the situation when a new platform (with an explicit OWL ontology instantiated) is to join this ecosystem. Here, a new "group of alignments" has to be created. Ontologies of all platforms that the incoming platform has to communicate with, need to be aligned with its ontology. In the mentioned example, if the new platform Q is to communicate bidirectionally with all present platforms, then eight new alignments have to be created. For the logistics scenario, let us assume that port management system, road haulier's platform, dynamic lighting platform, driver's mobile application platform already interconnected. A new truck owner's platform would like to join and communicate with all of them. The alignments between truck owner platform's ontology and ontologies of four other platforms would need to be created, with truck owner's ontology as a target or a source (depending on the direction of the required communication).

The advantage of the one-to-one alignments approach is a good quality of translations between ontologies, since the alignments are generated directly between ontologies of two systems, without any "intermediate ontology" (see, next section). Moreover, it is easy to understand and debug the translation process.

The main drawback of this approach is its poor scalability. When the ecosystem is mature and has many connected platforms, the new/incoming systems may need to connect to many of them, which results in preparation and storage of many alignments. As a result, the procedure for joining the ecosystem becomes complex, because incoming IoT platform administrator (semantic engineer) needs to obtain other platforms' ontologies, create alignments (using recommended tools) and optionally verify/modify/extend automatically created alignments. Finally, for obvious reasons, maintenance of such system becomes a nightmare, as changes in ontologies of each individual partner need to be tracked across multiple repositories of alignments (so that their effects are dealt with).

4.2 Central Ontology

The second possible approach is based on a central ontology that is created from *merged* IoT and domain ontologies. By domain ontology we understand conceptual model for specific domain, e.g. transportation. It should be controlled and maintained by the semantic engineer. This ontology is to be created when the initial ecosystem is put in place.

This approach is depicted in Fig. 2, where one ontology named GOIoTP (Generic Ontology of IoT Platform) constitutes the central point for achieving interoperability. Note that the informational scope of this ontology depends on the specific deployment environment. Let us again consider four platforms P, R, S, T. Here, an alignment in each direction is created between each platform's ontology and the GOIoTP ontology.

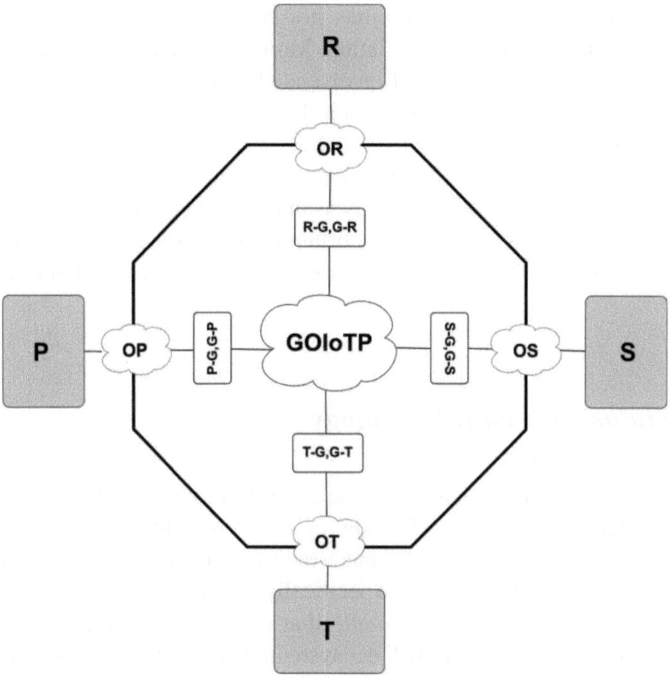

Fig. 2 Central ontology

Obviously, in a special case when one of the joining IoT systems is only a "feeder", it is possible to create only a "one-way" alignment, but such cases are likely to be rather rare and their existence does not change the overall structure of the discussed approach.

For example, let us consider the situation in which platform P wants to send a message to platform R (while each platform has its own knowledge representation model). The input data is represented in the OP ontology and, using the P-G alignment, it can be represented in the central ontology. Then, using the G-R it can be translated to the OR ontology and understood by platform R. When platform R wants to send a response, it is translated with the R-G to the central ontology, and then using the G-P to the the OP ontology so that it can be understood be the P platform.

Advantages of such an approach are many. First, it does not suffer from the scalability problems. Integration of new (incoming) platform involves creation of a single pair of alignments, with the central ontology. Second, it requires less preparation/work from programmer (semantic engineer) responsible for bringing the new IoT platform to the ecosystem. This is because only a single "point of joining" has to be instantiated. Furthermore, the long-term maintenance is simplified as changes in a single platform require localized adjustments only. However, the quality of translation may be worse that in case of one-to-one alignment since, for every messages traveling between two platforms, two alignments need to be applied. Some information that is platform specific may not be represented in the central ontology, and may be lost during the translation. Moreover, the problem arises when the central ontology has to be modified or extended, e.g. by adding new fragment of a domain ontology. This can happen when the joining application brings on board important, ut not represented, yet, in the GOIoTP aspects of the domain. In this case, all existing alignments have to be re-generated (as they can be negatively affected by the modification). While there are only as many such alignments as there are connected platforms, but this is still an important effort that is required. Furthermore, the resulting merged GOIoTP may give worse translation results than its previous version.

4.3 Modularized Central Ontology

The last approach is a modification of the central ontology approach. It addresses some of the identified problems. The key aspect of the proposed solution is based on the observation that, in order to make the central ontology less vulnerable to changes, it should be modular [7]. Specifically, the central—modular—ontology G should consist, first, of a *core module* that models knowledge common to all stakeholders participating in the IoT ecosystem. This could be an ontology of the Internet of Things (similar, for instance, to [41, 42]). It would then be complemented by domain specific ontologies that are context specific (e.g. a transport and logistics ontology, see [43]). Each of these ontologies would constitute separate module (in the central ontology G) and, possibly, be modular itself. Furthermore, modules brought to the ecosystem by the platforms would also become part of G.

Let us now consider what happens when a need to add new knowledge to the system (that has not been present so far) arises. This may be a result of a new "communication topic" that should be handled by the connected platforms, e.g. for the aforementioned scenario—meteorological conditions at the estimated time of truck arrival. In this case, an ontology extension(s) have to be included.

Observe also that individual modules (including extensions) do not have to be used by all parties connected to the ecosystem, but only platforms that are interested in this (specific) "topic" of communication. For the logistics example, let us recall that there is an ecosystem consisting of platforms representing port management, road haulier, container terminal, and truck owner. Truck owner and road haulier platforms may want to exchange information describing trucks technical details (e.g. engine type and/or engine power) that are not interesting to other partners. This information is also so unique that, it should not be modeled in a separate module belonging to G. Specifically, a new ontology module—a *Truck Specification Ontology*—can be designed and imported into the central ontology (G).

It should now become clear that the proposed approach, while similar to the one with a central ontology, allows different way of handling alignments. Alignments to/from a particular module of G will be generated only by the platforms that need it in their communication. In Fig. 3, within the central ontology G, four modules

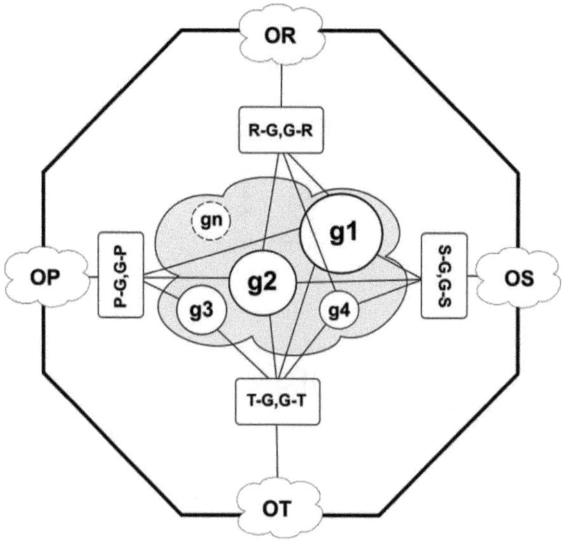

Fig. 3 Modularized central ontology

(g1, g2, g3, g4) are depicted. Modules g1 and g2 are used by all platforms, i.e. they contain general concepts such as IoT devices or basic truck identification descriptions. Module g3 models information specific to P and T, and module g4 for platforms R, S, T. Therefore, for instance, there will be alignments between ontology of P and module g3, but no alignments between module g3 and R or S. This also supports our decision to store alignments, which can then be used for faster translation of context specific messages (e.g. messages exchanged between P and T).

It should now be clear in what way the modularized central ontology is a solution positioned between the one-to-one alignments and the central ontology approaches. It inherits the easier method of creating/joining the ecosystems, since a smaller number of (specific communication targeted) alignments are created, while the central ontology is easier to manage. Since the alignments will be "context specific" there is a chance that the quality of translation will be better than in the case of the central-merged ontology.

5 Proposed Architecture

As shown above, our current approach to achieving semantic interoperability between IoT platforms is based on a *modularized central ontology*. Let us now discuss how we envision the structure and characteristics of an architecture utilizing this approach.

To address the problem of message translation the key component, is the *IoT Platform Semantic Mediator* (IPSM). Its core purpose is a semantic-driven translation of messages. In other words, the IPSM as its input will have messages expressed in

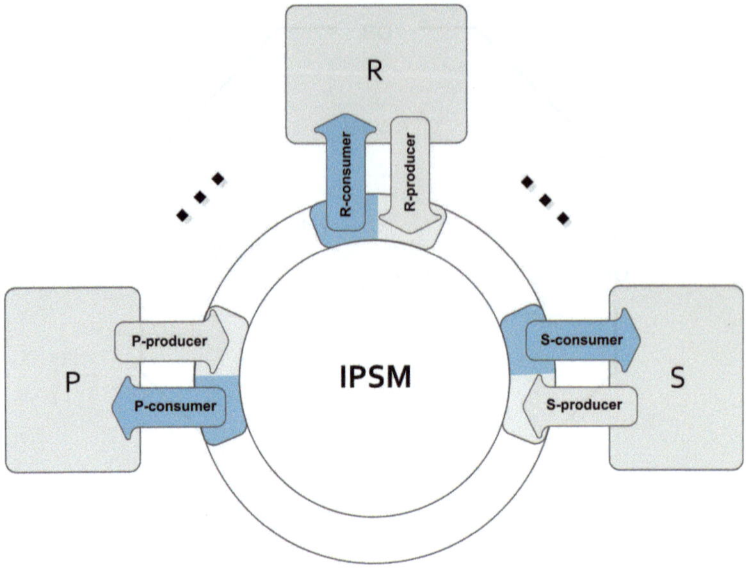

Fig. 4 IPSM instance: "federation" of platforms

terms of one platform and as its output will produce message expressed in terms of ontology of the second platform. It's important to stress that the IPSM is meant to be a generic module/component, deployed in many *instances*, each of them joining together a specific "federation" (set) of platforms. Each instance provides a semantic interoperability infrastructure, tailored towards the goals, for which given ecosystem has been formed. Its central point is the *modular* common ontology, denoted by G in Fig. 5, which (as it should become obvious by now) serves as a "lingua franca" in the translation process (Fig. 4).

Any platform P willing to participate in a "federation" using the IPSM instance needs a pair of alignments between its own ontology and the common ontology G. The "into G" and "from G" alignments are denoted by P => G and G => P in Fig. 5, and constitute the basis for the *P output* and *P input translator* respectively. We assume that the alignments will be provided/constructed by a semantic engineer, acting as *platform integrator*. The input/output translators are created by instantiating a generic *translator component* provided by IPSM.

It is worth observing that the IPSM does not need to "know" the complete ontology of any of the federated platforms—all that is needed for the translation are the alignments with the common ontology G. This design decision both simplifies the internal structure of the IPSM and allows the platforms to "reveal" only those parts of their internal ontologies that are necessary for communication purposes, i.e. for constructing the "translation alignments".

The IPSM is assumed to be completely external-data-format agnostic. In other words, its inputs and outputs are "typed" by ontologies of communicating platforms.

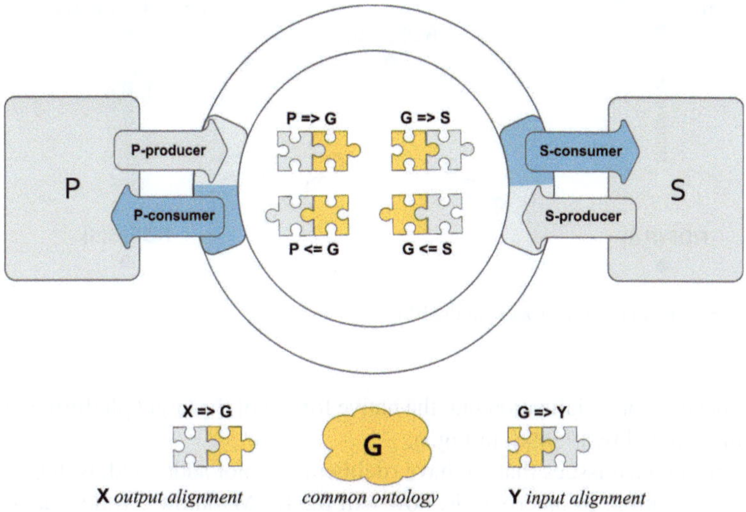

Fig. 5 IPSM internal architecture

However, there is one more conversion step that has to be performed. Let us assume that messages that a platform "emits" are expressed in XML. Then a *de-serialization* has to occur that will convert them into their semantic counterparts. Obviously, *serialization* is needed "on the way back". An important question arises—where will this procedure take place. An obvious answer seems to be that it is the "outside world" that has to perform the transformation of messages to/from the platform-specific format. In this case, any platform that is to join a common IoT ecosystem, in addition to the aforementioned alignments, needs two more crucial components—a *producer* and a *consumer* (cf. Fig. 4). The producer is responsible for de-serialization of platform-specific data into its *semantic representation*, whereas the consumer—for the inverse process. Both components need to be constructed based on the knowledge of the data serialization formats used by the platform and the platform semantics/ontology. The semantic representation of serialized data is expected to be given in the platform ontology, expressed in OWL, and wrapped-up in RDF triples. One might also envision an architecture in which the IPSM "includes" the producers/consumers, of course. Discussion of pros and cons of both approaches is out of scope of this contribution and thus is omitted. We just assume that the producer and consumer have been created and are available for the (de-)serialization process.

Let us now consider the flow of messages between two IPSM-connected platforms P and S, depicted in Fig. 6. For any P message d, the P-producer converts it to the RDF representation RDF(P,d) and "injects" into the IPSM. The process is facilitated by the IPSM API. The RDF data is further translated, using the alignments P => G and G => S. The application of the first alignment gives RDF(G,d), expressed in the common ontology G. The second alignment produces an RDF data RDF(S,d), which uses the ontology of the target platform S. The result is subsequently injected into

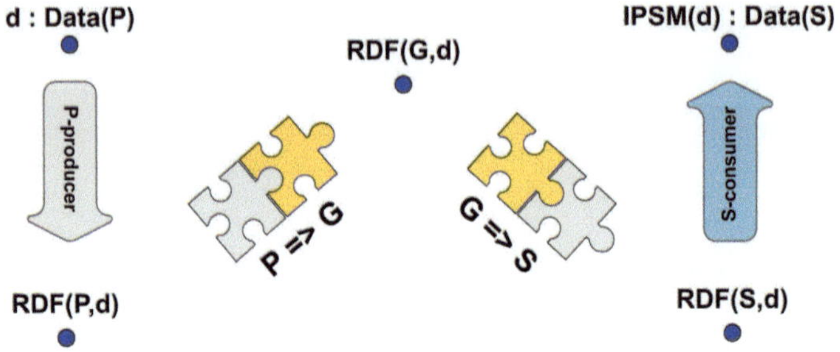

Fig. 6 Semantic translation process in the IPSM

the S-consumer for serialization into the native format of the target platform S, giving the result denoted by IPSM(d) in Fig. 6.

One of the open issues that we have (deliberately) not addressed, is the question of message routing. In other words, how will the IPSM "know", which "language" it should translate messages to, and where to send the result. Similarly, one can ask, how is the IPSM going to deal with more complex communication patterns, e.g. one to many broadcasting. While the most obvious answer is that it will receive additional metadata with the needed information, it is also possible that the IPSM will only deal with translations, while the "message routing" will be provided by the ecosystem in which it is to operate. However, addressing this issue is outside of scope of the current contribution.

Let us now summarize the key points of our proposed architecture. The IoT ecosystem can utilize an arbitrary number of IPSM instances, each of them serving a particular goal, and connecting a subset of potential platforms. The same platform can, of course, participate in several "interest groups", i.e. be connected to several IPSM instances. For a platform to communicate via a given IPSM instance, the following essential components are needed:

- *translation alignments* P => G and G => P, where G denotes the common ontology of the IPSM instance, and P—the ontology of the platform,
- *producer* and *consumer* modules, responsible for conversion between platform's native data format and its semantic representation.

The process of constructing the above components requires cooperation between the platform engineers and semantic engineers.

6 Concluding Remarks

The aim of this chapter was to propose a way to facilitate *semantic interoperability* between IoT artifacts (applications/systems/platforms). The proposed approach starts from lifting to OWL ontologies of all artifacts that are to be joined in an IoT ecosystem. Obviously, entities based on OWL ontologies do not need to be adjusted. Next a common modular ontology G is instantiated. It consists of a “core IoT ontology” and modules representing domains of communication within the ecosystem. In the following step, a set of alignments is created, forming the basis for the semantic translation. When message flows are known in advance (which platform communicates with which and what is the “topic of conversation”), only alignments that are needed for translation of expected messages can be created and stored. The IoT ecosystem may utilize one or more instances of the IPSM, which will be used for message translation. The translation process starts from a message sent in a format specific to a given platform. It is de-serialized (in the producer module) into the OWL-based representation. Next, the alignments are used for a “double translation”, first to the common ontology, then to the ontology of the receiver. Finally, the consumer module serializes the message into the data format specific to the receiving platform.

Since our work is a part of a much bigger project, a large number of open questions remain. The two that we have indicated here were. First, where in the ecosystem the producer and the consumer modules will be placed (and who will be responsible for their creation). The second, which is much more elaborate (and answer to which is likely to affect an answer to the first), concerns the architecture of the interoperability ecosystem. In other words, how the IPSM will be actually used within it, as far as message routing is concerned. However, it should be noted that answers to these questions are not going to affect the above described approach to semantic interoperability and the architecture of the IPSM.

Acknowledgements Research presented in this paper has been partially supported by EU-H2020-ICT grant INTER-IoT 687283.

References

1. Fortino, G., Trunfio, P. (eds.): Internet of Things Based on Smart Objects. Technology, Middleware and Applications. Springer, Berlin (2014)
2. Murdock, P., Bassbouss, L., Datta, S.K., et al.: Semantic interoperability for the Web of Things. Technical report EURECOM+5000, Eurecom, 09 (2016)
3. Staab, S., Studer, R.: Handbook on Ontologies, 2nd edn. Springer, Berlin (2009)
4. Owl 2 web ontology language profiles. https://www.w3.org/TR/2009/REC-owl2-profiles-20091027/
5. The Description Logic Handbook: Theory, Implementation and Applications, 2nd edn. Cambridge University Press, Cambridge (2010)
6. Owl 2 syntaxes. https://www.w3.org/TR/2012/REC-owl2-overview-20121211/#Syntaxes
7. Stuckenschmidt, H., Parent, C., Spaccapietra, S. (eds.): Modular Ontologies. Concepts, Theories and Techniques for Knowledge Modularization, volume 5445 of State-of-the-Art Survey, LNCS (2009)
8. Euzenat, J., Shvaiko, P.: Ontology Matching, 2nd edn. Springer, Berlin (2013)
9. Ganzha, M., Paprzycki, M., Pawłowski, W., Szmeja, P., Wasielewska, K., Fortino, G.: Tools for ontology matching—practical considerations from INTER-IoT perspective. In: Proceedings of the 8th International Conference on Internet and Distributed Computing Systems. LNCS, vol. 9864, pp. 296–307. Springer, Berlin (2016)

10. Otero-Cerdeira, L., Rodríguez-Martínez, F.J., Gómez-Rodríguez, A.: Ontology matching: a literature review. Expert Syst. Appl. **42**(2), 949–971 (2015)
11. Ganzha, M., Paprzycki, M., Pawłowski, W., Szmeja, P., Wasielewska, K., Palau, C.E.: From implicit semantics towards ontologies—practical considerations from the INTER-IoT perspective (submitted for publication). In: Proceedings of 1st edition of Globe-IoT 2017: Towards Global Interoperability among IoT Systems (2017)
12. Owl 2 web ontology language mapping to RDF graphs. https://www.w3.org/TR/owl2-mapping-to-rdf/
13. Bahrami, M., Hacherouf, M., Nait Bahloul, S.: Proceedings of the 2015 international conference on soft computing and software engineering (SCSE'15), DTD2OWL2: a new approach for the transformation of the DTD to OWL. Procedia Comput. Sci. **62**, 457–466 (2015)
14. Rodrigues, T., Rosa, P., Cardoso, J.: Mapping XML to existing OWL ontologies. In: Isaías, P., Nunes, M.B., Martínez, I.J. (eds.) International Conference WWW/Internet 2006, pp. 72–77 (2006)
15. Bohring, H., Auer, S.: Mapping XML to OWL ontologies. In: Leipziger Informatik-Tage, volume 72 of LNI, pp. 147–156. GI (2005)
16. Yahia, N., Mokhtar, S.A., Ahmed, A.: Automatic generation of OWL ontology from XML data source. *CoRR*, abs/1206.0570 (2012)
17. Bedini, I., Matheus, C., Patel-Schneider, P.F., Boran, A., Nguyen, B.: Transforming XML schema to OWL using patterns. In: 2011 Fifth IEEE International Conference on Semantic Computing (ICSC), pp. 102–109 (2011)
18. Ferdinand, M., Zirpins, C., Trastour, D.: Lifting XML schema to OWL. In: Koch, N., Fraternali, P., Wirsing, M. (eds.) Web Engineering - 4th International Conference, ICWE 2004, Munich, Germany, 26–30 July 2004, Proceedings, pp. 354–358. Springer, Heidelberg (2004)
19. Ghawi, R., Cullot, N.: Building ontologies from XML data sources. In: Proceedings of the 20th International Workshop on Database and Expert Systems Application, DEXA '09, pp. 480–484. IEEE Computer Society, Washington (2009)
20. Tsinaraki, C., Christodoulakis, S.: Interoperability of XML schema applications with OWL domain knowledge and semantic web tools, pp. 850–869. Springer, Berlin (2007)
21. Tsinaraki,C., Christodoulakis, S.: XS2OWL: a formal model and a system for enabling XML schema applications to interoperate with OWL-DL domain knowledge and semantic web tools, pp. 124–136. Springer, Berlin (2007)
22. JSON schema. http://json-schema.org/
23. Droettboom, M.: Understanding JSON Schema. https://spacetelescope.github.io/understanding-json-schema/
24. http://www.altova.com/xmlspy/json-schema-editor.html
25. Jsons2xsd. https://github.com/ethlo/jsons2xsd
26. Cullot, N., Ghawi, R., Yétongnon, K.: DB2OWL: a tool for automatic database-to-ontology mapping (2007)
27. Cerbah, F.: Learning highly structured semantic repositories from relational databases. In: European Semantic Web Conference, pp. 777–781. Springer, Berlin (2008)
28. Ontology alignment evaluation initiative. http://oaei.ontologymatching.org/
29. Jiménez-Ruiz, E., Grau, B.C.: LogMap: logic-based and scalable ontology matching. In: International Semantic Web Conference (ISWC). LNCS, vol. 7031, pp. 273–288. Springer, Berlin (2011)
30. Aumueller, D., Do, H.-H., Massmann, S., Rahm, E.: Schema and ontology matching with COMA++. In: Proceedings of the 2005 ACM SIGMOD International Conference on Management of Data, pp. 906–908. ACM (2005)
31. Faria, D., Pesquita, C., Santos, E., Palmonari, M., Cruz, I.F., Couto, F.M.: The AgreementMakerLight ontology matching system. In: On the Move to Meaningful Internet Systems: OTM 2013 Conferences, pp. 527–541. Springer, Berlin (2013)
32. A format for ontology alignment. http://alignapi.gforge.inria.fr/format.html
33. SKOS simple knowledge organization system reference. https://www.w3.org/TR/skos-reference/

34. Maedche, A., Motik, B., Silva, N., Volz, R.: MAFRA - a mapping framework for distributed ontologies. In: EKAW '02: Proceedings of the 13th International Conference on Knowledge Engineering and Knowledge Management, Ontologies and the Semantic Web, pp. 235–250, London, UK. Springer, Berlin (2011)
35. Web Ontology Language. https://www.w3.org/TR/owl2-overview/
36. Bouquet, P., Giunchiglia, F., van Harmelen, F., Serafini, L., Stuckenschmidt, H.: C-owl: contextualizing ontologies. In: Proceedings of the 2nd International Semantic Web Conference (ISWC 2003) (Sanibel Island, Florida). LNCS, vol. 2870, pp. 164–179. Springer, Berlin (2003)
37. SWRL: A semantic web rule language combining owl and ruleML. https://www.w3.org/Submission/SWRL/
38. RIF overview. https://www.w3.org/TR/rif-overview/
39. EDOAL: Expressive and declarative ontology alignment language. http://alignapi.gforge.inria.fr/edoal.html
40. David, J., Euzenat, J., Scharffe, F., Trojahn dos Santos, C.: The alignment API 4.0. Semant. Web **2**(1), 3–10 (2011)
41. IoT lite Ontology. http://www.w3.org/Submission/iot-lite/
42. SSN Ontology. http://www.w3.org/2005/Incubator/ssn/ssnx/ssn
43. Ganzha, M., Paprzycki, M., Pawłowski, W., Szmeja, P., Wasielewska, K.: Towards Semantic Interoperability Between Internet of Things Platforms (submitted for publication). Springer, Berlin (2016)

Linked Data for Internet of Everything

Danh Le-Phuoc and Manfred Hauswirth

Abstract The Internet and the World Wide Web have transformed modern life by connecting the real world. With the massive connectivity extension of the Net to a much wider world wide web of the manifold physical objects, we are on the verge of the next evolution of the Internet, so called the Internet of Everything (IoE). However, enabling seamless interoperability is still a biggest challenge towards building the next generation of IoE applications. Believing in Linked Data as a promising solution to address this challenge, we propose the idea of "Linking Everything" by extending Linked Data Principles to interlink "everything" into a hypergraph. Via this book chapter, we present how to make this hypergraph programmable via the emerging Semantic Web technologies.

1 Introduction

The Internet and the World Wide Web have transformed modern life by connecting the real world. With the massive connectivity extension of the Net to a much wider world wide web of the manifold physical objects, we are on the verge of the next evolution of the Internet, so called the Internet of Everything. Hence,

"Cisco defines the Internet of Everything (IoE) as bringing together people, process, data, and things to make ***networked connections more relevant and valuable than ever before-turning information into actions*** *that create new capabilities, richer experiences, and unprecedented economic opportunity for businesses, individuals, and countries."*

D. Le-Phuoc (✉) · M. Hauswirth
Technical University of Berlin, Berlin, Germany
e-mail: danh.lephuoc@tu-berlin.de

M. Hauswirth
e-mail: manfred.hauswirth@tu-berlin.de

R. Gravina et al. (eds.), *Integration, Interconnection, and Interoperability of IoT Systems*, Internet of Things, DOI 10.1007/978-3-319-61300-0_7

The Internet makes it easy to develop networked applications, but there are too many competing non-interoperable networking technologies and there will be a lot more to come. Therefore, the ambition of turning data generated from things across system boundaries suffers from a lack of interoperability across platforms. As a result, developers are facing with data silos, high costs and limited market potentials. Consequently, every programmer must understand the methods available to retrieve data from each API, and write custom code for accessing data from each data source. This might lead to an explosion in small, specialized applications (or mashups) that combine data from several sources, each of which is accessed through an API specific to a data provider.

To remedy this interoperability chaos, we advocate for employing Linked Data Principles [5] based on RDF[1] (Resource Description Framework) to build platform independent APIs for application developers. The Linked Data Principles are a killer solution to solve the problem of lacking reliable onward links signposting the way to related data which usually comes as isolated fragments. Hence, to have effective utilization of data across systems, Linked Data Principles make data linkable and therefore discoverable via simple content negotiation processes. On top that, RDF provides a flexible way to describe things in the world such as people, things, process, data or abstract concepts and how they relate to other things. These statements of relationships between things are, in essence, links connecting things in the world. As a consequence, such links between items in different data sources connect them into a single global information space.

Hence, in this book chapter we introduce the idea of *"Linking Everything"* into a global hypergraph based on Linked Data Principles. Different from the Web architecture, we propose the extensions to the Linked Data Principles for web data sources based on HTTP(s) to adopt to a much wider range of networking technologies and software stacks in the Internet of Things (IoT) domain [12]. However, to lower the effort for data consumers to discover, access and integrate data as Linked Data for the web does, the simple access methods using standardized data structures, i.e., serialized RDF formats is still the key element of our proposal. Such access methods aim to make it agile for third parties to consume and work with the data without a prior agreement with the publishers as well as precedent knowledge of data sources.

1.1 Running Example

To illustrate the whole idea of the book chapter, this section will introduce a running example that shows different aspects of integrating data generated from things into several application domains regardless network technologies and IoT platforms. This example includes various data integration demands at run-time from everything that can provide useful data to build smart applications. The first type of exemplar applications is home automation which needs to interact and gather data from home

[1] https://www.w3.org/RDF/.

appliances such as smart fridges, smart ovens, thermostats, Phillips Hue Lights. The second type of application domain is Smart Cities which provides traffic data, weather data and also smart infrastructures/APIs for smart car/navigations. There is a need for these two application domains to integrate each other on the fly. For example, the smart cities platform could provide the accurate estimation time to arrive home based on the current traffic situation so that a home automation platform will know when to turn on thermostats or warm up smart ovens.

The cross-platform integration demand is likely to be expanded other domains. For instance, with respect to smart shopping domain, the phone can talk to the smart cities platform to find out nearby retail stores which sell the products in a user's shopping list stored in her/his phone. Then, the phone can query the prices from potential stores to compare the prices to decide which one to buy it and pick up in the most convenient way (fastest route or without a detour) based on the current traffic condition. To enable this, the smart city platform, the smart navigation platform and the smart shopping platform have to enable real-time discovery and interaction without any pre-scripted application scenarios.

1.2 Outline

To realize our proposal of using Linked Data to enable the vision of Internet of Everything, we will introduce an extended version of Linked Data principles for interlinking physical things and virtual things in the next section. These principles will provide a foundation for a data fusion strategy to link everything into a hypergraph in the following Sect. 3. Next, Sect. 4 will introduce the enabling technologies to publish, access and query the proposed hypergraph. Finally, the Sect. 5 will conclude the book chapter.

2 Linking Everything by Extending Linked Data Principles

To realize the vision of interlinking Everything using Linked Data, we extend the Linked Data principles [5] that are widely adopted to interlink data across web-based information systems into a global information space on top of the World Wide Web general architecture. Next, we will discuss in details how such principles will be used and extended according to the ordered listed in below.

1. Use URIs as names for things
2. Use HTTP URIs so that people can look up those names.
3. When someone looks up a URI, provide useful information, using the standards (RDF, SPARQL)
4. Include links to other URIs. So that they can discover more things.

2.1 Using More Than HTTP URIs for Naming Things

To be able to interlink physical objects connected to the Internet via different IT systems/infrastructures, such physical objects need to be globally identifiable to transparently correlate different pieces of information about them across system boundaries. For example, a smart home gateway can connect to the manufacturer's portal to get the specification of its smart fridge and the mobile phone can access to that smart home gateway to query information of the fridge. Similarly, information of a food product spreading from its manufacturer to retail stores or online shopping websites, then delivery systems to a smart home, or specifically, to a smart fridge can be integrated in a single mobile application. To this end, following the first principle of Linked Data, using URIs as global references for physical objects is a natural way to enable the global identification for physical objects.

However, as opposed to the recommended practice of Linked Data on the Web that advocates to use only HTTP URIs [5], other URI schemes should be used to cover the pervasiveness of IoE. In particular, due to mobility nature of physical things, they can not be constantly connected to a specific network to be accessible via a permanent HTTP URI. For instance, the phone can share the owner's favourite playlist to a car entertainment system or a TV set top box via an ad-hoc network in a car or in a smart home. In essense, CoAP protocol can be used instead of the traditional HTTP with a web server, e.g., coap://myphone.local/ for CoAP over UDP or coap+tcp://myfridge.local/ over TCP. Along this line, a non-protocol-binding URI naming scheme like EPC[2] which is designed to assign universal identifiers for every physical object anywhere in the world can be used for uniquely naming things. In particular, a UPC (Universal Product Code) or an EAN(International Article Number) barcode which are widely used in retail stores or online shop can be encoded in URI notations of EPC. For example, the UPC-A bar code 616368417123 of a milk box can be encoded in the URI form as **urn:upc-a:616368417123**. The identifier of this form is the anchor to tap into various valuable online information of a product, for example, from nutrition facts to packing history or online offers of similar products.

Note that a physical object can have multiple URIs as its aliases. In a certain application context, a subset of its alias are linked together to reflect the contextual knowledge (see Identity links in Sect. 2.4) about that object. For example, an LG NEXUS have three URIs, **urn:upc-a:652810119627**, **urn:imei:353490068949375** and **urn:asin:B00IDSYK2A**. Two URIs, **urn:upc-a:652810119627**, **urn:imei: 353490068949375** are encoded from UPC-A barcode and IMEI[3], number assigned by LG. Within LG's database, LG knows that IMEI number 353490068949375 registered to a GSM network is associated with the phone tagged with the UPC-A code, 652810119627. When the phone is delivered to Amazon, Amazon assigns an

[2]EPC, https://en.wikipedia.org/wiki/Electronic_Product_Code.

[3]IMEI, https://en.wikipedia.org/wiki/International_Mobile_Station_Equipment_Identity.

ASIN number,[4] B00IDSYK2A, for its internal tracking. With two URIs, **urn:upc-a:652810119627** and **urn:asin:B00IDSYK2A**, Amazon can link information of this phone in its inventory management system to the product portal of the manufacturer LG. A smart home application running on this phone can assign another URI such as **coap://myphone.local/** to interface with other home appliances but still can use the URI encoded from UPC-A bar code to search for firmware/software updates from LG.

Along with the URIs to identify a tangible real world object, other relevant URIs can be generated to reference to data items generated by it. Furthermore, such URIs for data items might be associated with the URIs referencing to abstract concepts defining the semantics and relationships among those data items. For example, a thermostat can generate temperature readings which can be assigned with URIs. The URIs referencing to such readings can be linked with the abstract concept, **Room Temperature** associated with a URI which is defined in an ontology. A group of such URIs can be encompassed in an online resource made accessible for a certain range of applications, e.g., the resource *http://mythermostat.local/temperature/* contains recent temperate reading(s) and relevant information. However, the temperature readings change in realtime, instead of such a pull-based resource using HTTP, it can be replaced by a push-based resource via a CoAP endpoint based on Websocket, e.g., *coap+ws://mythermostat.local/temperature/*.

2.2 Looking Up/Interacting with Things Via Multiple Protocol Bindings

Follow the second principle of Linked Data which uses HTTP URIs as the resources to point to data items in the domain of interest, each resource in the form of HTTP URI can return data items associated with properties and relationships as a Web document. While the HTTP protocol is a universal access mechanism on the Web that normally requires a Web server to serve Web documents as resources. However, HTTP is based on TCP protocol using point-to-point communication model that is not suitable for push-based notification services and probably too complex for some constrained devices. This leads to the development of other similar protocols that are more light-weight but still able to carry out similar interaction patterns of HTTP. For example, different from HTTP, CoAP can operate over UDP instead of using a complex congestion control as in TCP [30, 31]. However, CoAP is based on REST architecture, which is a general design for accessing Internet resources in the same way of the HTTP content negotiation.

The basic idea of content negotiation is that HTTP/CoAP clients send HTTP/-CoAP headers with each request to indicate what kinds of documents they prefer.

[4]The Amazon Standard Identification Number (ASIN) is a 10-character alphanumeric unique identifier assigned by Amazon.com and its partners for product identification within the Amazon organization.

Servers can inspect these headers and select an appropriate response. If the headers indicate that the client prefers JSON-LD, then the server will respond by sending a JSON document. If the client prefers Turtle, then the server will send the client a Turtle document. A server is intended to be an autonomous data provider without prior knowledge about the client's capabilities, thus, it can serve different types of standardised data formats so that it is accessible for various types of clients (browser, resource-constraint devices, etc.). For example, following is the negotiation processes of via HTTP and CoAP following the W3C Linked Data Platform standard for HTTP [34] and COAP [23] respectively. In this example, HTTP binding can provide the resource description of *the Phillip Hue sensor in the living room* under two formats, Turtle and JSON-LD, while the COAP one only can return Turtle. Note that, both HTTP and COAP resources can coexist if the server (e.g., smart home gateway) can support both of them via its two networking interfaces.

Listing 1 Content Negotiation via HTTP

```
GET /phillipHue01 HTTP/1.1
Host: mysmarthome.local
Accept: text/turtle

HTTP/1.1 200 OK
Content-Type: text/turtle; charset=UTF-8
Link: <http://www.w3.org/ns/ldp#BasicContainer> rel="type",
      <http://www.w3.org/ns/ldp#Resource> rel="type"
Allow: OPTIONS,HEAD,GET,POST,PUT,PATCH
Accept-Post: text/turtle, application/ld+json
Accept-Patch: text/ldpatch
Content-Length: 250
ETag: W/'123456789'

@prefix ssn: <http://www.w3.org/ns/ssn/>.
@prefix sh: <http://www.smarthome.org/ns/>.
@prefix ldp: <http://www.w3.org/ns/ldp#>.
@prefix: <http://mysmarthome.local/>.

<:phillipHue01> a ldp:Container, ldp:BasicContainer, a :AimbientLightSensor;
   ssn:madeObservation: light:2016:09:25:08:04:50, :light:2016:09:25:08:04:51;
   sh:locatedIn :livingRoom.
```

Listing 2 Content Negotiation via COAP

```
GET coap://mysmarthome.local/phillipHue01
Accept: text/turtle

2.05 Content
Content-Format (ct): text/turtle
ETag: W/'123456789'


@prefix ssn: <http://www.w3.org/ns/ssn/>.
@prefix sh: <http://www.smarthome.org/ns/>.
@prefix ldp: <http://www.w3.org/ns/ldp#>.
@prefix: <coap://mysmarthome.local/>.

<:phillipHue01> a ldp:Container, ldp:BasicContainer, a :AimbientLightSensor;
   ssn:madeObservation :light:2016:09:25:08:04:50, :light:2016:09:25:08:04:51;
   sh:locatedIn :livingRoom.
```

Look closely to the above content negotiation sequences and patterns of HTTP and COAP, they can be implemented on top other existing protocols. The popular ones are the widely used protocols in home automation such as BACnet[5] for controlling lights, blinds and room units, Echonet[6] for controlling ACs. Note that some of them have IP-based implementations, but, several of them can work over lower network layers like layer 4 or layer 2 of the OSI model.[7] For example, while Echonet is built on top of layer 4, BACNet can work over the data link/physical layer, including ARCNET, Ethernet, BACnet/IP, Point-To-Point over RS-232, Master-Slave/Token-Passing over RS-485, and LonTalk . On top that, lower power consumption technologies such as BLE[8] and Zigbee[9] can provide a sufficient network transportation stack to emulate such content negotiation features,e.g, GAP[10] and GATT[11] APIs.

Based on such a content negotiation mechanism, the resource will provide the client a document in a format that the client is able to parse and understand the content. This document plays the role as a container comprising information about the object that the data provider would like to expose to the consumer. The common practice is that the URIs identifying real-world objects are different from URIs for such documents describing them. This practice allows separate sets of statements to be made about an object and about a document describing that object. This mechanism is quite useful for bootstrapping the following processes of the interaction between two devices exchanging information, i.e. following the links returned in each step of the interactions. Therefore, two devices can carry out the interactions/exchanges via different communication channels accessible for them. For example, a mobile can discover a service endpoint via BLE Physcial Web beacons[12] from which the mobile can then established a secured HTTPS channel via its LTE network.

2.3 Dynamic RDF Information Retrieval

Following the third Linked Data principle advocating the use of a unified data model, Resource Description Framework (RDF), for publishing structured data, we propose to extend this principle towards representing the dynamic nature of data generated in the information sphere around real world things. RDF is a simple graph-based data model that has been designed for Semantic Web to link up web documents under controlled vocabularies and semantics. With RDF, we can represent structured data using sentences similarly to what we do in using natural language to express facts in

[5]BACnet protocol, https://en.wikipedia.org/wiki/BACnet.

[6]Echonet Protocol https://echonet.jp/about_en/.

[7]https://en.wikipedia.org/wiki/OSI_model.

[8]https://en.wikipedia.org/wiki/Bluetooth_low_energy.

[9]https://en.wikipedia.org/wiki/ZigBee.

[10]GAP REST API White Paper, https://www.bluetooth.org/docman/handlers/downloaddoc.ashx?doc_id=285911r.

[11]GATT REST API White Paper, https://www.bluetooth.org/docman/handlers/downloaddoc.ashx?doc_id=285910.

[12]https://google.github.io/physical-web/.

daily life. In essence, RDF enables the use of signs, syntax, and semantics to encode and decode information (data in some context).

The underlying model (or schema) of RDF (and other languages) is first-order logic, i.e., the observation-of-fact (data) that is an expression of the notion that everything is related to something else, in a variety of ways. Using RDF, a system can expose anything at any resource hosted in an entry point bound to any networking protocols in a manner that is comprehensible for both machine and human. The below example represents a set of related sentences about a Phillips Hue light sensors. e.g., *":phillipHue01 is an ambient light sensor which observes photon luminance property of the living room"* and *" sensor :phillipHue01 made an observation:light:2016:09:25:08:04:50 which has the result with the value of 100 at 2016-09-25T08:04:50"*

Listing 3 RDF sentences

```
:phillipHue01 a cf:AimbientLightSensor;
            ssn:observes qu:photonLuminance;
            ssn:madeObservation:light:2016:09:25:08:04:50.
:light:2016:09:25:08:04:50 ssn:observedProperty qu:photonLuminance;
                    ssn:hasResult [ssn:hasValue 100^xsd:float;
                         ssn:resultTime 2016-09-25T08:04:50^xsd:dateTime].
qu:photonLuminance ssn:isPropertyOf :livingRoom.
```

This presentation is a text-based standard serialization format of RDF, Turtle. RDF can also be serialized in other standard formats like RDFXML and JSON-LD. Each above sentence is composed of RDF triples which are distinguished by two types, literal triples and RDF links. A literal triple has an RDF literal such as a string, number, or date as the object. Literal triples are used to describe the properties of resources. For instance, literal triples are used to express the value of a light sensor reading of a sensor observation. Literals may be plain or typed, e.g., float or datetime. RDF links describe the relationship between two resources. RDF links consist of three URI references. The URIs in the subject and the object position of the link identify the related resources. The URI in the predicate position defines the type of relationship between the resources. For instance, the last triple in above example specifies the relationship between the living room and the property photon luminance that the light sensor is observing.

A useful distinction can be made between internal and external RDF links. Internal RDF links connect the resources within a single Linked Data source. Thus, the subject and object URIs are in the same namespace. External RDF links connect resources that are served by different Linked Data sources. The subject and object URIs of external RDF links are in different namespaces. External RDF links are crucial for interlinking things across the system boundaries as they are the glue that connects data islands into a global, interconnected data space. Intuitively, this global data space is connected by a node-and-arc-labeled directed graph as an integrated representation of information that originates from multiple sources, is heterogeneously structured, and is represented using different schemata [5].

To access to data to this global graph, one can start from any node as an RDF resource to traverse through the RDF links connecting this node to the adjacent RDF nodes using the content negotiation mechanism described in the above Sect. 2.1.

A more convenient and powerful way to retrieve a subset of nodes and arcs in this graph is using the standard query language for querying RDF graphs, SPARQL 1.1.[13] SPARQL 1.1 standard defines a HTTP-based protocol to enable a HTTP client to query the graph via a HTTP content negotiation using a SPARQL query in the text-based format, called SPARQL endpoint. Hence, a SPARQL endpoint can be adopted to other non-HTTP protocol like CoAP to give access to any RDF graph hosted in an embedded gateway. On top of that, SPARQL 1.1 enables the query federation on top a set of SPARQL endpoints that enables a powerful mechanism to integrate RDF graphs stored in distributed systems (more technical details discussed in Sect. 4.2.1)

Note that, in listing 3, the light sensor continuously generates light observations like the set of triples linked to *:light:2016:09:25:08:04:50*. Therefore, such data arrives to the system as a stream, called Linked Data Stream or RDF stream [22]. To get the fresh data on such RDF streams, the client has to continuously pull the RDF resources or pose the query to the SPARQL endpoint. This pull-based fashion of retrieving data from RDF streams is inefficient due to the real-time nature of the streaming data source, e.g., sensory data. Therefore a push-based fashion should be used to reactively retrieve data similar to pub/sub systems whereby the data consumer subscribes to a streaming channel via Websocket or MQTT protocol and then when the streaming sources have the data, it will push to the consumer, e.g., TripleWave [24]. Also, to be able to continuous query RDF streams, some extensions of SPARQL 1.1 are proposed to enable reactive queries over streaming data sources (more technical details discussed in Sect. 4.2.2)

2.4 Interlinking Things at Realtime

The fourth Linked Data principle advocates the use of hyperlinks to connect any types of things. For example, a hyperlink may be set between *"a light sensor"* and *"my living room"*, or between a property of interest, *"Photon Luminance"* and a place, *"my living room"*. Note that, such hyperlinks connecting things have types which describe the relationships between the things. Such types are defined in ontologies, e.g. the relationships defined in listing 3 are defined in SSN ontology [8]. These statements of relationships between things are, in essence, links connecting things in the world. As a result, such data items are equipped with reliable onward links signposting the way to related data [5]. Therefore, publishing links among everything using Linked data principles make them linkable and therefore discoverable.

The first type of RDF link is Identify Link. Identity Links point at URI aliases used by other data sources to identify the same real-world object or abstract concept. Identity links enable clients to retrieve further descriptions about an entity from other data sources. Note that in the context of multiple protocol bindings, one physical object can have different URIs associated with the protocols they expose data to. Also, Identity links have an important social function as they enable different views

[13] https://www.w3.org/TR/sparql11-query/.

of the world to be expressed on the Web of Data. For instance, a food product can have different URIs assigned by manufacturers, shipping companies and the retail stores. Each URI reflects what each publisher knows about a product in a document which includes Identity links like rdfs:seeAlso or owl:sameAs [14].

The next type of RDF links is Vocabulary Link. Vocabulary Links point from data to the definitions of the vocabulary terms that are used to represent the data, as well as from these definitions to the definitions of related terms in other vocabularies. Vocabulary links make data self-descriptive and enable Linked Data applications to understand and integrate data across vocabularies. Such self-descriptive information enables the client to discover new data sources by following RDF links at run-time but also to help them to integrate data from these sources. Integrating data requires bridging between the schemata that are used by different data sources to publish their data. The term schema is understood in the Linked Data context as the mixture of distinct terms from different RDF vocabularies that are used by a data source to publish data on the Web. This mixture may include terms from widely used vocabularies as well as proprietary terms.

The last type of RDF links that needs significant extensions to adapt dynamic and mobile nature of IoT are Relationship Links. Relationship Links point at related things in other data sources, for instance, places and DBpedia are hosted as background databases of things on the Web [5]. However, in IoT, such relationships are often dynamic relationship links are only formed in certain time and space context, e.g., a device is connected to a certain network or a thing is nearby other things. The dynamic relationship links are generated in realtime that will create transient information routes/paths among information spaces. For instance, when a car contacts a city infrastructure to query traffic status, there is an information path between its routing path and integrated data on traffic status (traffic lights, accident reports and road works).

Moreover, algorithmic relationship links are functional RDF predicates that are used to present some certain built-in functions in the RDF query processing engines. The spatial computation functions are the first type that can use the links of this kind to map to geometric computations supported in a RDF store. For example, the predicates like *geo:nearby*, *geo:within* can be used to map respective functions to compute over RDF resources that have spatial properties. The desired features of RDF processing engines are inferring implicit facts, i.e., RDF triples from other RDF triples generated from raw data, e.g., sensor readings. With the supports of RDFS, OWL or RIF reasoners, we can express some RDF predicates which can be mapped to a set of reasoning rules based on the ground RDF facts. Going towards applying machine learning to generate symbolic facts as RDF statements which can then be used as RDF links inside graph patterns of SPARQL, a SPARQL query engine can support some predefined predicates representing built-in functions which can trigger a classification or clustering algorithm for asserting the desired facts. For example, based on a computer vision algorithm, it can classify if a traffic jam at a junction based on traffic camera readings. A triple pattern {*?jucntion :isTrafficJam True*} in a SPARQL query can trigger that algorithm to assert this type of RDF statements.

3 Agile Data Fusion with a Unified HyperGraph of Everything

With the proposed extended principles that use RDF as a unified data model for integration across IoT data silos, all data represented as RDF graphs will be fused into a hypergraph. As a graph-based model agnostic to IoT protocols, software stacks, application domains and infrastructures, the hypergraph offers an agile way to incrementally link up anything at real-time without a precedent knowledge of data schema and platforms. Following is the gradual steps of fusing data generated by everything into such a hypergraph. For more technical details, we would like to refer readers to our work, Graph of Things [21].

3.1 Physical Sensors

The first step of fusing data about physical things is deploying physical sensors to *observe the properties of interest* of such real world things. The RDF graph pattern generated by a sensor is driven by the Semantic Sensor Network ontology [8] that provides four interlinked perspectives of a sensing activity on a real world thing. The core perspective is the observation data and related metadata. Then, next is feature and property perspective, focusing on what senses a particular property or what observations have been made about a property. Also these two perspectives are linked to sensor perspective, with a focus on what senses, how it senses, and what is sensed; and system perspective, with a focus on systems of sensors and deployments.

Figure 1 is an example RDF data layout driven by SSN ontology. The figure gives a visual view of how sensor observation data of a light sensor and a temperature sensor in a living room are connected into as a subgraph of the global hypergraph of everything, called Linked Sensor Data. The upper layer is the sensor metadata showing sensor perspective and system perspective, i.e., temperature sensor and light sensor observe temperature property and photon luminance of the living room respectively (serialized form of this subgraph is presented in listing 3).

The lower layer of the Fig. 1 shows a snapshot of the sensor observations of both sensors in the same time point *2016-09-25T08:04:50*. Note that a stream slice in the lower layer is a subgraph connected into the hypergraph at realtime. The stream slice is linked to upper layer via two links: ssn:isPropertyOf and ssn:observedProperty which show which properties the observations were made for. The Turtle form of such a subgraph is shown in below.

Listing 4 Sensor observations in Turtle

```
:phillipHue01 cf:AimbientLightSensor;
        ssn:madeObservation:light:2016:09:25:08:04:50,:light:2016:09:25:08:04:51, :light:2016:09:25:08:04:52.
:light:2016:09:25:08:04:50:observedProperty qu:photonLuminance;
                    :hasResult [:hasValue 100^xsd:float; :resultTime 2016−09−25T08:04:50^xsd:dateTime];
:thermostat01 cf:TemperatureSensor;
        ssn:madeObservation:temp:2016:09:25:08:04:50, :temp:2016:09:25:08:04:51, :temp:2016:09:25:08:04:52.
:temp:2016:09:25:08:04:50 :observedProperty qu:photonLuminance;
                    :hasResult [:hasValue 200^xsd:float; :resultTime 2016−09−25T08:04:50^xsd:dateTime];
```

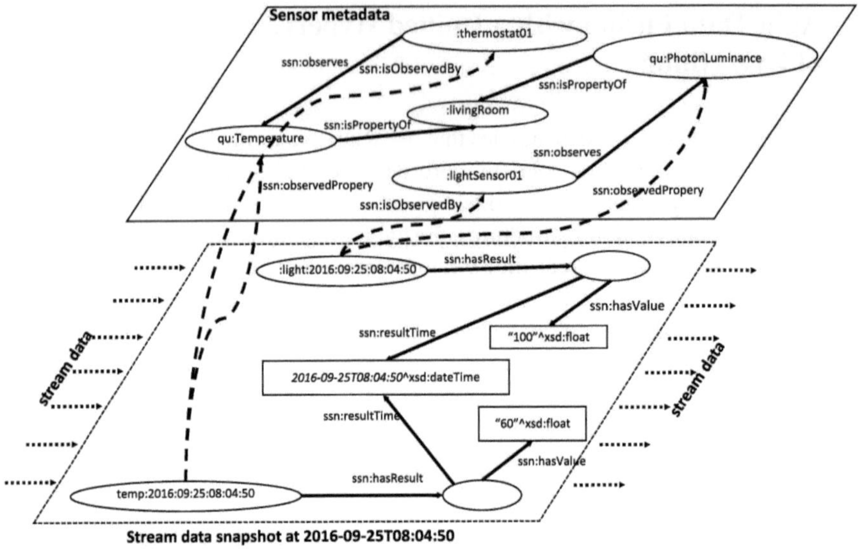

Fig. 1 Linked sensor data

3.2 *Citizen Sensors*

Humans continuously subconsciously and consciously sense, process, and induce inferences from events around them in real-time. Sense in this context is defined as one of the methods for a living being to gather data about the world: sight, smell hearing, touch, and taste. Humans also leverage past experiences, background knowledge, and reasoning to extract meaning from often confusing or new experiences [17]. These roles show that the human sensors perform processing and analysis of the collected data [33]. This pre-processing/processing of sensory data from experience/background knowledge is what differentiates our sensing capabilities from hardware sensors.

As a user of the Web/Internet, human reports his/her observations via Web platforms in textual form in natural language [32] like writing blog, sending text, or microblog. Therefore, humans as citizens on the ubiquitous Web, acting as sensors and sharing their observations and views using mobile devices and Web 2.0/3.0 services. Twitter is a representative example of microblogging platform which can see as a social sensing "device". As microblogging lends itself to instantaneous updates, creation of data related to events around the world is posted before it can be reported on by more traditional media methods or even by blog or blog-like services.

Sheth [32] envisages that microblogging platforms like Twitter (where users can post short textual messages, Internet links, attached videos, and pictures) as low-effort publishing services are of particular interest to citizen sensing due to the large scale of Twitter users, which enables the user to post from mobile devices with minimum effort through mobile applications [17]. Twitter posts and Instagram feeds created by web citizens using mobile devices report observations of these events in almost real-time. However, such social observations are the composition of human observations and physical sensing observations (e.g., GPS coordinates, photos, videos). Such physical observations are instrumented by sensing devices mount in mobile phones own by the human. The combination of both human as a sensor and physical sensors annotating human observation is defined as Citizen Sensor [13, 32] or probably with other names, urban sensing or participatory sensing [9, 10, 18].

3.3 Knowledge-Based Sensors

The observation and measurement are the low-level data that is captured by sensors, other devices or human users. This could be large volume of data related to an entity of interest or an environment. However, the IoT data consumers (i.e. users and applications) are often interested in the high-level concepts that refer to machine-interpretable or human-understandable knowledge. A sample application of creating such high-level abstractions is discussed in [3, 15] where sensory observation data is used in a logical inference model to derive perceptions from the raw observations. The data abstraction and knowledge extraction processes to enable transforming low-level IoT data to high-level knowledge that refer to an event, a pattern, are comprehensible to the machines and human users, and play an important role in leveraging the full potential of IoT.

These events or patterns are spatially, temporally, and thematically linked [32]. These events and series of related events can rapidly create networks of users posting information which are related to background knowledge bases such as geospatial database (LinkedGeo dataset or GeoNames), DBpedia, Google Knowledge Graph. The correlations among observations generated by physical sensors and human sensors are by driven the knowledge bases that are curated and annotated by domain experts, so called knowledge-based sensors. The curation and annotation process led by domain experts would enhance the disseminating of raw observations to the high level abstract observations such as events, situations or symbolic statuses driven by semantics.

For example, Tweets can be used conduct data analytics in combination with knowledge bases for inferring the correlation of space and time to identify events [16]. In the process, a citizen sensing platform first might need to gather topically relevant data by keyword-based search in Twitter. This works on the presumption that high volume search terms describe events and are of high interest to users [27]. Then, to categorize retrieved Tweets, it need to correlate with a taxonomy of hashtag which might be curated or modelled as an ontology designed by a domain expert.

4 Enabling Technologies

4.1 Exposing Dynamic Linked Data

4.1.1 Publishing Dynamic Linked Data for Things

To publish data for Things following the extended Linked Data Principles in Sect. 2, the standards and software stacks [5] having developed for publishing Linked Data on the Web are the crucial foundations for adding new recipes to adapt the dynamic nature of IoE. The first thing to extend is enabling data publication not only on HTTP as HTTP might be too heavy for many resource-constraint devices which are the dominant citizens of IoE. However, the best practice is emulating the necessary features of HTTP on top other protocols to have consistent and transparent interaction pattern between data consumer and publisher. For example, in [23], a CoAP-based implementation is provided to conform Linked Data Platform standard [34] which is originally created for publishing RDF resources on top of HTTP.

While CoAP operates over UDP, typically in IPv6 environments, CoAP extends the concept of client-server Internet interaction to the micro-controller level. Unlike HTTP, CoAP has an inbuilt mechanism for observing resources, allowing clients to request server to push updates in real-time. Therefore, CoAP is more suitable for streaming data in the context where bandwidth and battery and computing resource are scarce. CoAP is a one-to-one protocol for transferring state information between client and server. While it has support for observing RDF resources, CoAP is best suited to a state transfer model like HATEOAS,[14] e.g., CoRE Link [31] is a RESTful instance of CoAP.

For publishing the resources that can change in real-time in a high frequency, MQTT might be a better option to CoAP as CoAP is not purely event-based. Moreover, in terms of implementation, MQTT has a simpler protocol specification even both MQTT and CoAP are suitable for implementation in sensor devices typically using 8–32 bits microcontrollers with 200 MHz clock rates and 8K-64M RAM. CoAP clients are simpler than HTTP, but more complex than MQTT. In terms of data transport, MQTT employs connection oriented communication given by TCP, where CoAP uses cheaper UDP. The use of TCP means more data exchanged than in UDP (TCP header is 20 bytes vs 8 bytes in UDP).

For a setting that needs to support variety of protocols including HTTP, Websocket, CoAP and MQTT, Hypercat standard[15] provides an open, lightweight JSON-based hypermedia catalogue format for exposing collections of URIs. Each HyperCat catalogue may expose any number of URIs, each with any number of RDF-like triple statements about it. HyperCat is simple to work with and allows developers to publish linked-data descriptions of resources. HyperCat is primarily designed for exposing information about IoT assets over the web. Hypercat catalogues are JSON documents,

[14] https://en.wikipedia.org/wiki/HATEOAS.

[15] http://www.hypercat.io/standard.html.

typically served over HTTP(S). However, a Hypercat catalogue may list any number of resources by URIs of any type, accessed with any protocol. CoAP has a fully documented URI scheme which may be used in Hypercat documents. Also, MQTT URLs may be embedded in Hyper MQTT URLs may be embedded in Hypercat documents. In general, Hypercat allows a server to provide a set of resources to a client, each with a set of semantic annotations. Implementers are free to choose or invent any set of annotations to suit their needs. Hypercat rels may be declared to inform clients of the kind of data to expect from an MQTT stream. For example, subscribing to the provided topic may provide payloads in a particular MIME-type.

4.1.2 Data Enrichment

As highlighted in Sect. 2.4, adding RDF links is building bridges connecting data islands. It is important noting that providing RDF links alone does not provide semantic interoperability and will not solve all the issues regarding discovery, management of data, and supporting autonomous interactions. The RDF links as semantic description still needs to be shared, processed, and interpreted by various methods and services across different domains. Therefore, such RDF links have to be driven by standardized ontology like SSN or commonly used like www.schema.org. Such ontologies play the role as the common understandings on the vocabularies defining typed RDF links and controlled semantics.

For the current and existing applications, it is also important that their ontologies and knowledge bases can be accessed and reused by large groups of potential consumers. Developing and sharing ontologies and contributing towards description and annotation frameworks that can support legacy applications are effective steps in achieving semantic interoperability on a large scale. Other solutions, such as ontology mapping and matching (i.e. manual, semi-automated, or automated) can help linking the resources using different semantic annotation models. The ontology designers can also reference existing common ontologies and provide links to other upper-level ontologies to support interoperability between different semantic descriptions in the IoT domain [3].

Due to the realtime and mobility nature of Things, temporal and spatial context should be captured along with sensory data or connectivity status. Therefore, temporal and spatial annotations using controlled vocabularies being standardized by W3C Working Group, Spatial Data on the Web,[16] will facilitate spatial-temporal computation operators for the consumption parties. For instance, location of a car will play the role of the parameter to discover near by traffic sensors like traffic cameras, congestion detectors and road works.

[16]https://www.w3.org/2015/spatial/.

4.2 RDF Data Processing

The vision of "Linking Everything" into the Internet will create a network of decentralized RDF Graphs. From the infrastructure perspective, there will be a federated setup whereby the RDF Data Processing engines can distributedly coordinate the processing to carry out a data processing task across heterogeneous data sources bound to various networking protocols. Due to the heterogeneity of the data sources and infrastructures that serve them, there is a wide range of capabilities of RDF Query Engines are needed to realize the vision of Internet of Everything. Next, we will introduce some standard features of RDF query engines that are commonly used by IoT applications. Then, more advanced features such as continuous queries and reasoning will be followed.

4.2.1 Standard RDF Query Engines

There has been a large of body of work in building RDF Query Engines, e.g., Hexastore [36], TripleBit [37] and RDF-3X [28] and Sesame [7], Jena TDB [25], RDFox [26] and OWLIM [4]. Among them, the ones supporting SPARQL 1.1 like Jena, Virtuoso and Stardog provide desired features of various IoT-based data integration scenarios. For example, the following query is used to correlate the my shopping list with a geo-spatial endpoint and other SPARQL endpoints of retailed stores nearby my current location.

```
SELECT DISTINCT ?item ?price ?store.

WHERE {

       SERVICE <?storeEndpoint>{
                 ?item schema:price ?price.
       }
       SERVICE <geo:endpoint>{
                 ?mylocation geo:nearby ?store.
                 ?store sparql:endpoint ?storeEndpoint.
       }
       GRAPH <myphone:shoppinglist>{
                  []      :outofStock ?item
       }
       GRAPH <myphone:gps>{
                  :mycar     :hasLocation ?location
       }
}
```

In above query, there is lot of machineries working under the hood. The basic operation is the basic graph pattern sub-queries over RDF graphs stored in local RDF stores, probably on the smartphone, where the local the query engine sits, e.g. *myphone:shoppingList* and *myphone:gps*. Similar sub-queries are relayed over a remote SPARQL endpoint to remotely execute there. To coordinate the processing with the federation services (via SERVICE keyword) of SPARQL 1.1, the query executor has to interleave the internal processing together with remote executions from other endpoints. Note that, there is a nested discovery that pipelines the outputs from *geo:endpoint* with the spatial computation via the functional predicates *geo:nearby* (Cf. Sect. 2.4) via the variable *?storeEndpoint* which points to the SPARQL endpoints of potential stores.

4.2.2 Stream Data and Continuous Queries

Traditionally, above standard RDF query engines only support pull-based query based on SPARQL 1.1. However, sensory data is often represented as streams and is distributed over different networks with diverse types of data. As the data is real-time and the attributes of data (i.e. quality attributes) can change over time, the query mechanisms for IoT need to address this dynamicity and agility [3]. For example, in IoT, resources can appear and disappear over time, the data can be collected from different heterogeneous resources, and real time processing of data streams is required for event detection.

In recent years, there have been an increasing number of work on supporting stream processing or complex event processing using RDF data model. These bodies of work are motivated by sensory data generated from different IoT domains. Most of them extend standard functionalities in SPARQL 1.1 to support querying features of streaming data. The stSPARQL and stRDF [19] extend the SPARQL query language and RDF representations with spatial and temporal dimensions to facilitate query on sensor data which is mostly time and location dependent. C-SPARQL [2], streaming SPARQL [6] and CQELS [20] are other extensions of the SPARQL query language to support continuous queries over streaming data. EP-SPARQL [1] is an extension to SPARQL that enables processing complex events and stream reasoning. Following is an example continuous query extended from SPARQL 1.1 grammar, called CQELS-QL. This query continuously correlates two RDF streams, a stream on accidents happening on the road and a stream of car locations from the car's GPS navigation. Besides, the OPTIONAL keyword will correlate the camera footage nearest to the accidents' location if available.

```
prefix spatial: <http://rdfs.org/sioc/ns/spatial/>

SELECT DISTINCT ?acclocation ?mylocation ?distance ?camlocation

WHERE {
       STREAM  <city:accidents>{
               ?accident :reportedAt ?acclocation. }
       STREAM <:gps>[NOW]{
               :mycar :hasLocation ?mylocation.
```

```
            :mycar :currentRoute ?myroute.
            :acclocation spatial:within ?myroute.}
    OPTIONAL{
            STREAM <city:trafficcameras>[2 minutes]{
                    ?acclocation spatial:nearest ?camlocation:
                    ?camerafeed :hasLocation ?camlocation. }
    }
}
```

4.2.3 Reasoning

Above RDF query processing engines can be equipped with reasoning capabilities by offering an inference layer on top of the query layer, for instances, Sesame [7], Jena TDB [25], RDFox [26] and OWLIM [4]. There are several classes of reasoning capabilities which enable different expressivity levels of logic rules. Light weight RDFS and OWL rules enable to express several useful subsumptions which are can be used to infer transitive relationships in SPARQL, especially, abstracted types. This types of light weight logical rules can be efficiently handled by recent reasoning engines like RDFox [26] and OWLIM [4] and Inferray [35] with the scale of billions of RDF triples.

In terms of lightweight subsets of RDFS and OWL, approaches for forward-chaining are classified into two categories based on their underlying algorithmic approaches. the RETE Algorithm [11] and iterative rules systems. The RETE algorithm works on pattern matching principles where rule sets are represented as a DAG and are matched to facts that are represented as relational data tuples. The current version of the algorithm is not public due to intellectual property constraints. However, several systems have implemented their own flavor of RETE [25]. Furthermore, RETE has recently been ported to GPU architectures [29]. Iterative rule systems offer a simpler design. Rules are iteratively applied to the data until a stopping criterion is matched. These systems usually operate on a fixed-point iteration principle, i.e., the inference process terminates when an iteration (application of all rules) derives no triples. Similarly, the OWLIM reasoners family uses a custom rule entailment process with a fixed-point inference mechanism. However, dedicated hardware is still required compared to the usual commodity servers. Targeting on mainstream architectures, Infferay [35] offer an efficient reasoner efficiently run on any commodity machine. To overcome the memory bandwidth bottleneck, it focuses on designing a data structure and algorithms that favor sequential access.

5 Conclusions

This book chapter addresses the interoperability issue of Internet of Everything by proposing the idea of “Linking Everything”. The idea is realized by laying out the extended version of Linked Data Principles for publishing and consuming data on the Web. The adopted principles for Internet of Everything pave the way to interlink everything into a hypergraph which is made programmable agnostic to networking technologies and platforms. The book chapter leverages the emerging Semantic Web technologies as the key enablers for seamless data intergration across system boundaries and IoT platforms.

References

1. Anicic, D., Fodor, P., Rudolph, S., Stojanovic, N.: Ep-sparql: A unified language for event processing and stream reasoning. In: Proceedings of the 20th International Conference on World Wide Web, WWW '11, pp. 635–644. ACM, New York, NY, USA (2011)
2. Barbieri, D.F., Braga, D., Ceri, S., Grossniklaus, M.: An execution environment for c-sparql queries. In: Proceedings of the 13th International Conference on Extending Database Technology, EDBT '10, pp. 441–452. ACM, New York, NY, USA (2010)
3. Barnaghi, P., Wang, W., Henson, C., Taylor, K.: Semantics for the internet of things: early progress and back to the future. Int. J. Semant. Web Inf. Syst. **8**(1), 1–21 (2012)
4. Bishop, B., Kiryakov, A., Ognyanoff, D., Peikov, I., Tashev, Z., Velkov, R.: Owlim: a family of scalable semantic repositories. Semant. web **2**(1), 33–42 (2011)
5. Bizer, C.: Evolving the web into a global data space. In: Proceedings of the 28th British National Conference on Advances in Databases, BNCOD'11, pp. 1–1. Springer, Berlin (2011)
6. Bolles, A., Grawunder, M., Jacobi, J.: Streaming sparql extending sparql to process data streams. In: Proceedings of the 5th European Semantic Web Conference on The Semantic Web: Research and Applications, ESWC'08, pp. 448–462. Springer, Berlin (2008)
7. Broekstra, J., Kampman, A.: Harmelen, F.v.: Sesame: a generic architecture for storing and querying RDF and RDF schema. In: Proceedings of the First International Semantic Web Conference on The Semantic Web, ISWC '02, pp. 54–68. Springer, London (2002)
8. Compton, M., Barnaghi, P., Bermudez, L., García-Castro, R., Corcho, O., Cox, S., Graybeal, J., Hauswirth, M., Henson, C., Herzog, A., Huang, V., Janowicz, K., Kelsey, W.D., Le Phuoc, D., Lefort, L., Leggieri, M., Neuhaus, H., Nikolov, A., Page, K., Passant, A., Sheth, A., Taylor, K.: The SSN ontology of the w3c semantic sensor network incubator group. Web Semant. **17**(C), 25–32 (2012)
9. Cuff, D., Hansen, M., Kang, J.: Urban sensing: out of the woods. Commun. ACM **51** (2008)
10. Demirbas, M., Bayir, M.A., Akcora, C.G., Yilmaz, Y.S., Ferhatosmanoglu, H.: Crowd-Sourced Sensing and Collaboration Using Twitter (2010)
11. Forgy, C.L.: Expert systems. In: Chapter Rete: A Fast Algorithm for the Many Pattern/Many Object Pattern Match Problem, pp. 324–341. IEEE Computer Society Press, Los Alamitos, CA, USA (1990)
12. Fortino, G., Trunfio, P.: Internet of Things Based on Smart Objects: Technology, Middleware and Applications. Springer Publishing Company, Incorporated (2014)
13. Goodchild, M.: Citizens as sensors: the world of volunteered geography. GeoJournal **69** (2007)
14. Halpin, H., Hayes, P.J., McCusker, J.P., McGuinness, D.L., Thompson, H.S.: When owl: Sameas isn't the same: an analysis of identity in linked data. In: Proceedings of the 9th International Semantic Web Conference on The Semantic Web - Volume Part I, ISWC'10, pp. 305–320. Springer, Berlin (2010)
15. Henson, C., Sheth, A., Thirunarayan, K.: Semantic perception: converting sensory observations to abstractions. IEEE Internet Comput. **16**(2), 26–34 (2012)
16. Janowicz, K.: The role of space and time for knowledge organization on the semantic web. Semant. Web **1**(1, 2), 25–32 (2010)
17. Kamel Boulos, M.N., Resch, B., Crowley, D.N., Breslin, J.G., Sohn, G., Burtner, R., Pike, W.A., Jezierski, E., Chuang, K.Y.S.: Crowdsourcing, citizen sensing and sensor web technologies for

public and environmental health surveillance and crisis management: trends, OGC standards and application examples. Int. J. Health Geograph. **10**(1), 1–29 (2011)
18. Kanhere, S.S.: Participatory Sensing: Crowdsourcing Data from Mobile Smartphones in Urban Spaces (2011). doi:10.1109/MDM.2011.16. http://dx.doi.org/10.1109/MDM.2011.16
19. Koubarakis, M., Kyzirakos, K.: Modeling and querying metadata in the semantic sensor web: the model strdf and the query language stsparql. In: Proceedings of the 7th International Conference on The Semantic Web: Research and Applications, ESWC'10, vol. Part I, pp. 425–439. Springer, Berlin (2010)
20. Le-Phuoc, D., Dao-Tran, M., Parreira, J.X., Hauswirth, M.: A native and adaptive approach for unified processing of linked streams and linked data. In: Proceedings of the 10th International Conference on The Semantic Web, ISWC'11, vol. Part I, pp. 370–388. Springer, Berlin (2011)
21. Le-Phuoc, D., Nguyen Mau Quoc, H., Ngo Quoc, H., Tran Nhat, T., Hauswirth, M.: The graph of things. Web Semant. **37**(C), 25–35 (2016)
22. Le-Phuoc, D., Xavier Parreira, J., Hauswirth, M.: Linked Stream Data Processing, pp. 245–289. Springer, Berlin (2012)
23. Loseto, G., Ieva, S., Gramegna, F., Ruta, M., Scioscia, F., Sciascio, E.D.: Linked data (in resourceless) platforms: a mapping for constrained application protocol. In: Proceedings of the 15th International Conference on The Semantic Web, ISWC'16. Springer, Berlin (2016)
24. Mauri, A., Calbimonte, J.P., Dell'Aglio, D., Balduini, M., Brambilla, M., Della Valle, E., Aberer, K.: Triplewave: Spreading RDF streams on the web. In: The Semantic Web - ISWC 2016: 15th International Semantic Web Conference. Kobe, Japan, October 17–21, 2016, Proceedings, Part II, pp. 140–149. Springer, International Publishing, Cham (2016)
25. McBride, B.: Jena: Implementing the RDF model and syntax specification. In: Proceedings of the Second International Conference on Semantic Web -. SemWeb'01, vol. 40, pp. 23–28. CEUR-WS.org, Aachen, Germany (2001)
26. Motik, B., Nenov, Y., Piro, R., Horrocks, I., Olteanu, D.: Parallel materialisation of datalog programs in centralised, main-memory rdf systems. In: Proceedings of the Twenty-Eighth AAAI Conference on Artificial Intelligence, AAAI'14, pp. 129–137. AAAI Press (2014)
27. Nagarajan, M., Gomadam, K., Sheth, A.P., Ranabahu, A., Mutharaju, R., Jadhav, A.: Spatio-Temporal-Thematic Analysis of Citizen Sensor Data: Challenges and Experiences, pp. 539–553. Springer, Berlin (2009)
28. Neumann, T., Weikum, G.: Rdf-3x: a risc-style engine for RDF. Proc. VLDB Endow. **1**(1), 647–659 (2008)
29. Peters, M., Sachweh, S., Zündorf, A.: Large scale rule-based reasoning using a laptop. In: Proceedings of the 12th European Semantic Web Conference on The Semantic Web. Latest Advances and New Domains, vol. 9088, pp. 104–118. Springer Inc, New York (2015)
30. Shelby, Z.: The constrained application protocol (coap). RFC 7252 (proposed standard) (Jun 2014). http://www.ietf.org/rfc/rfc7252.txt. Accessed 17 Sep 2016
31. Shelby, Z.: Constrained restful environments (core) link format. RFC 6690 (proposed standard. http://www.ietf.org/rfc/rfc6690.txt. Accessed 17 Sep 2016
32. Sheth, A.: Citizen sensing, social signals, and enriching human experience. IEEE Internet Comput. **13** (2009). doi:10.1109/MIC.2009.77
33. Srivastava, M., Abdelzaher, T., Szymanski, B.: Human-centric sensing. Philos. Trans. R. Soc. Lond. A Math. Phys. Eng. Sci. **370**(1958), 176–197 (2011). doi:10.1098/rsta.2011.0244
34. Steve Speicher John Arwe, A.M.: Linked data platform implementation conformance report. w3c working group note, w3c (Dec 2014). http://www.w3.org/TR/ldp-implreport/. Accessed 17 Sep 2016
35. Subercaze, J., Gravier, C., Chevalier, J., Laforest, F.: Inferray: fast in-memory RDF inference. PVLDB **9**(6), 468–479 (2016)
36. Weiss, C., Karras, P., Bernstein, A.: Hexastore: sextuple indexing for semantic web data management. Proc. VLDB Endow. **1**(1), 1008–1019 (2008)
37. Yuan, P., Liu, P., Wu, B., Jin, H., Zhang, W., Liu, L.: Triplebit: a fast and compact system for large scale RDF data. Proc. VLDB Endow. **6**(7), 517–528 (2013)

Sensing Enabled Capabilities for Access Control Management

Mikel Uriarte, Oscar López, Jordi Blasi, Oscar Lázaro, Alicia González, Iván Prada, Eneko Olivares, Carlos E. Palau, Miguel A. Portugués and Alejandro García

Abstract Current knowledge and assets that support organizations competitiveness must be protected. This protection is highly dependent on a proper access control management. Unfortunately, traditional access control management approaches are rigid and isolated, constrained by proprietary requirements not easily interoperable. In this chapter, a new framework approach is presented and described. It provides a flexible, open, fluid and collaborative middleware for building access control management systems, based on the Sensing Enriched Access Control (SEAC) concept. This framework establishes the principles allowing the development of an access

---

M. Uriarte · O. López · J. Blasi
Nextel S.A., Zamudio, Spain
e-mail: muriarte@nextel.es

O. López
e-mail: olopez@nextel.es

J. Blasi
e-mail: jblasi@nextel.es

O. Lázaro · A. González · I. Prada
Innovalia Association, Bilbao, Spain
e-mail: olazaro@innovalia.org

A. González
e-mail: agonzalez@innovalia.org

I. Prada
e-mail: iprada@innovalia.org

E. Olivares · C.E. Palau (✉)
Universidad Politécnica de Valencia, Valencia, Spain
e-mail: cpalau@dcom.upv.es

E. Olivares
e-mail: enolgor@dcom.upv.es

M.A. Portugués · A. García
Infoport Valencia S.A., Valencia, Spain
e-mail: maportugues@infoportvalencia.es

A. García
e-mail: agserrano@infoportvalencia.es

R. Gravina et al. (eds.), *Integration, Interconnection, and Interoperability of IoT Systems*, Internet of Things, DOI 10.1007/978-3-319-61300-0_8

control management system that copes with today organization's needs. The chapter also provides a description of a real use case raised to validate the framework, as well as the laboratory results supporting its scalability.

1 Introduction

The new economy, conditioned by the global competence under which organizations are, is strongly based on knowledge and asset management which must be protected from unauthorized access and preserved as key aspects for business continuity and success. Among other security aspects, one initial and relevant step involves access control management, granting access uniquely to right people and systems, both at physical and logical layers. Many (big) companies provide specific cards to their employee in order to be identified and be granted access to their premises. Even in the same buildings special rooms have specific access among the staff depending on the tasks and responsibilities to be accomplished. External visitors have to register in advance to be granted access for a limited period of time in a restricted area. In general terms there seems to be a natural trend to monitor any task operation performed by the staff or even systems, and thus access control plays a primary role to detect online and offline anomalies preserving a fluid performance of the company.

Unfortunately, access control has been typically developed in a very isolated approach, including legacy systems and specific particularities for each company solution. Even within the same company, different departments may develop different access control mechanisms. As a result, it is sometimes difficult to interoperate between systems and companies. As long as the number of companies implementing access control mechanisms is reduced, some particular connectors can be established between systems. However this approach is not scalable and sustainable as the number of interconnecting systems increases. The Internet of Things (IoT) era with billions of connected devices has arrived bringing new technologies and interaction ways, also covering security and thus the access control dimension. Interoperability is mandatory and some convergence must be provided in different domains and levels of the communication between entities (Things). Privacy and security have been identified as highly relevant challenges for IoT, so this chapter focuses on security in general and access control in particular.

It is important to note that sometimes security staff has been managed or divided into two groups, those responsible of the information security and those who are in charge of assuring physical security. This separation leads sometimes to ineffective situations due to a lack of interoperability. This is a fundamental reason why there is still much effort and study placed on security integration aspects, not only dealing with current legacy systems but also with emerging standards and paradigms.

Access Control (AC) is one of the main strategies for network security prevention and protection. Its main task is to fully share system resources, manage user's access rights and ensure that network resources are not used from unauthorized access. Unfortunately, traditional AC solutions are commonly an obstacle or, at least, are not properly adapted to new technologies such as the enterprise cloud, cloud-based apps, social media, and high-powered mobile devices which offer more ways to access corporate data and therefore more security issues. In the so called Bring Your Own Device (BYOD) era, moving data across various devices and networks increases security risks to the corporate network and may expose sensitive corporate data to leaks and attacks. This is because employee-owned mobile devices are beyond the scope of internal controls, and the growth in mobile malware brings new threats into the arena. There are some approaches to reduce risks, such as blacklisting and Bring Your Own Application (BYOA), which separates corporate and personal data on mobile devices using Mobile Application Management (MAM). Even so, security experts still point out that the ability to manage and track corporate data has become more difficult with the adoption of both cloud and mobile storage services in the enterprise. For that reason, nowadays organizations need to go beyond traditional security practices and policies and look at new ways for setting policies, control access and preventing data loss at both the application (logical) and device (physical) levels.

The BYOA concept relates to a partially local concept involving a user and its related company whereas the general perspective is much more complicated as we all currently live in the IoT, with a huge increase in the use of devices connected to the Internet (e.g. smartphones or smart objects). IoT also poses special attention to security aspects from a physical (device) and logical (cloud) level. Both levels add new security threats to those infrastructures where they are located, especially if they are Critical Infrastructures (CIs). Some clear examples can be found in smart meters or in connected cars or trucks. The huge amount of information generated by all these devices can be used to enhance business processes but also to improve access control in the different organisations [1].

If multiple companies with different devices have to interact in the IoT, there is a clear problem regarding interoperability for access control information exchange between organisations, which leads to implementing federated access control systems. Different proposals have been addressed to cover such problem; in our design we propose a framework that merges many pieces of information treating some of them as sensor data that serves as input for the required policies. This is called Sensing Enriched Access Control (SEAC) and will be further described in the next sections. In order to facilitate sensor interoperability, the management of sensor data is performed by using the Open Geospatial Consortium (OGC) founded Sensor Web Enablement (SWE) set of standards [2, 3]. The initiative promotes interoperability, defining various services and components, with a specific focus in geolocation. Positioning information allows to further extend the Role Based Access Control (RBAC) to geoRBAC. There are also other RBAC approaches that will be described in the next section.

Our SEAC framework includes a reasoning entity based on a CEP (Complex Event Processing) which allows further processing in order to apply access control policies using information from different sources or providers [2, 4]. The CEP can be either centralized for general policies, or it can be implemented locally for reduced environments due to critical or scalability reasons.

The proposed SEAC framework has been tested in the premises of a Critical Infrastructure (CI). CIs are logical/physical facilities that are essential assets in the economy and social well-being of a nation, and their incapacity or destruction may have a great impact in everyday life, debilitating security, national economy, national health or safety or a combination of different aspects. Therefore it is obvious that CIs are not static but dynamic and identified and classified in terms of impact and depend on their environment, e.g., a damaged base station in the city of New York has not the same impact in communication capacity as the same base station in a remote region in Africa. Generic examples of CIs include electrical power plants and the national electrical grid; supply services like food, fuel, water and gas; transportation systems like roads, railways, ports or airports; industrial control systems; and banking and financial services.[1] The testbed associated to the framework described in this chapter took place in the city of Valencia, and selected as CI the port, which is the sixth most important one in Europe and the second one in the Mediterranean see in terms of container traffic. Thus it turns out that Valencia's port is a CI at local, national and international level, so its choice is relevant as different parties involved in port transactions are interested in reliable access control solutions and may even provide access to a great number of different sensors to be integrated in our framework.

Here it is obvious the relevance of providing efficient and effective access control mechanisms and therefore we propose the use of our SEAC framework to further extend the current systems available in the port of Valencia. It includes various legacy physical systems and a Port Community System (PCS) as the main logical system. Basically, a PCS is a way of centralizing and coordinating the communication among all involved stakeholders within the port, involving both authority and transportation chain actors. Our proposed access control mechanism leverages the information available in the PCS, obtains positioning information from transportation companies, and highlights the need of interoperability.[2]

The current chapter exposes the ACIO (Access Control in Organizations) access control management framework and is structured as follows. First, some related work is presented regarding access control with particular attention in CIs in order to provide a general overview of current status, technologies and challenges. In the next section the concepts of Sensing Enriched Access Control (SEAC) and the role of IoT are further introduced and explained. Later on, the ACIO access control management framework is presented. Its main functioning principles, components and characteristics are detailed. The following section deals with the use case which serves to demonstrate an operational environment, overtaking existing deficiencies of the current access control management by means of the proposed solution based in the ACIO framework We will highlight the advantages provided by the ACIO components and its value for the access control management. Finally, the last section concludes by summarizing results reached throughout previous sections and presents further work (Table 1).

---

[1]Different countries have different CI classification, the European Union uses the one available https://ec.europa.eu/energy/en/topics/infrastructure/protection-critical-infrastructure.

[2]Available at http://www.portofrotterdam.com/.

Table 1 Most relevant European ports (source: port of Rotterdam)

Rank	Port	Volume 2014 (Million TEU)
1	Rotterdam, Holland	12.298
2	Hamburg, Germany	9.729
3	Antwerp, Belgium	8.978
4	Bremen, Germany	5.796
5	Algeciras Bay, Spain	4.555
6	Valencia, Spain	4.442

2 Access Control Overview

Access control mechanisms have been classified typically into physical and logical ones. Physical access control is defined as the security measures designed to avoid non-authorized access to specific premises, equipment or resources, and the protection of the people and personal from any harm or damage. Physical access control requires the definition of protection zones depending on their sensitivity or related risks. On the other hand, logical access control is related with the virtual environment in order to prevent non-authorized access to logical resources, data or services. However, with the increase in user mobility, global disposal of resources and services and the support that smart objects may require from virtual objects, physical and logical access control are being combined in order to create strong security measures in CI protection.

Physical Access Control mechanisms depend on the required security levels and have been classified in: (i) anti-vandalism measures; (ii) dissuasive measures; (iii) intrusion detection methods; and (iv) electronic surveillance. These mechanisms do not work isolated and are usually managed from a central control point which defines: (i) identity management; (ii) security policy management; and (iii) remote resources management. However, the main drawback of physical access control is the lack of interoperability and the interaction with logical access control mechanisms currently existing and under operation in different CIs.

With the current growth in the use of cloud services, collaboration between organizations and needed usability requirements, logical access control is evolving fast and new interoperability mechanisms are being defined. Processes like Single-Sign-On in which a user is authorized to access different resources with a single identification instance are broadly used. Moreover, administrators even delegate authentication in third parties like in cloud service access, e.g. SaaS (Software as a Service) or IaaS (Infrastructure as a Service). Some CIs, such as ports or logistic hubs, present an additional characteristic related with authentication and access control federation [5].

Identification and authentication are basic components for access control and different standards have been provided to cover different aspects. OpenID standardizes the authentication process; OAUTH 2.0 and SAML are related with the authorization framework whereas XACML is used as a policy definition language [6–9].

There are several standards that can be used for access control. The main difference between them is the mechanisms to manage user privileges. Mandatory Access Control (MAC), limits the access of the users depending on security attributes defined by authorization rules (policies) where only the administrator of the system defines the permissions that authorize each user to perform different actions over each object based on the attributes of both, in contrast with the Discretionary Access Control (DAC) which permits the transference of these privileges to the users in a discretionary way [10].

A more functional approach is found in the Role Based Access Control (RBAC): which considers all access permissions associated to roles, offering another abstraction layer to business level avoiding the need for user level granularity in the security policies, making it easier the maintenance of such policies in comparison to the other models (MAC and DAC) [10]. Attribute-Based Access Control (ABAC), in contrast to the previous models, specifies permissions based on attributes assigned to the subject, resource and environment, and only if the subject has the required attribute will be granted access to an object in a specific environment, this model is very effective in open systems as there is no need for the user to identify but to have the required attribute [11].

For dynamic and heterogeneous web services, we find that Semantic-aware Attribute-Based Access Control (SABAC) permits accessing to services based on attributes of related entities, and uses the Shibboleth service to address sensible attribute spreading, bringing the advantages of the semantic approach and reasoning capacity of emerging web applications that are missing in ABAC [12]. There is another version known as Semantic Context Based Access Control that combines semantic technologies with context based access control mechanisms mostly used in ubiquitous applications, such as mobile environments [13]. Other models manage permissions considering previous experience about past actions besides user rights over the object and environment, this is the case of the EBAM model (Experience Based Access Management) [14].

Another completely different approach is the Incentive Based Access Control (IBAM) model, where every subject is assigned a Budget for risks that has to be spent in order to obtain special rights over particular objects in extraordinary situations, using complementary mechanisms to control and reduce unnecessary risks [15]. For an IoT environment, Capability Based Access Control (CAPBAC) offers advantages for distributed management, support for delegation, access traceability, authentication chains to extend scalability and support for standard certificates based on Elliptic Curves (ECC) [16].

All these models manage access control under different perspectives and provide many grades of usability and security, some models are adequate to be implemented in systems that require high degree of security but not usability (MAC), and some others are designed to provide more usability even at the expense of a reduction in security (RBAC). The main problem that actual models face is that they focus access control as a dichotomy between security and usability, reducing one of those when the other one is reinforced, and that makes them not suitable for environments that require both.

3 The ACIO Framework

The ACIO framework is a compendium of principles, schemes and concepts that guide in the development of an open, flexible, scalable and highly secure and usable access control management system. In this section we will describe the Sensing Enriched Access Control concept, the ACIO framework principles and finally the proposed architecture.

3.1 Sensing Enriched Access Control (SEAC)

Access control principles are stated and immobile; there are three main options to identify an individual into a system: Who he is, what he has or what he knows. Examples of these three methods are biometrics (who he is), identity cards (what he has) and secrets, such as user and passwords (what he knows). Current and future improvements in the management of the previous principles will deal therefore with how this information is provided, stored, acquired and processed.

The Internet of Things (IoT) is a concept that deals with the connectivity capabilities enabled in physical objects, making them able to communicate over networks, such as the Internet. Most research carried out, in which security and IoT are present as keywords, deal with security requirements of this type of communications [17, 18]. However, in our previous bibliographic research and to the best of our knowledge the use of networked things contributing to the AC granting process in order to enhance security has not been deeply studied until the moment.

IoT enables the addition of multiple layers of information that can enhance access control decision making, such as presence sensors, biometric readers, productivity monitoring mechanisms, etc. The communication capabilities of such devices open the door to their inclusion as information into the equation of access granting, enabling the merge between two previously isolated management processes, physical and logical access control. This concept of including sensing information gathered through the sensor devices deployed in companies, organizations or CIs is what we identified as SEAC (Sensing Enriched Access Control). The idea by itself is not new, as there are context brokers able to gather sensor information; the innovation lies in its applicability to AC.

Under the scope of the SEAC concept, a sensor does not only limit to (physical) devices devoted to measure physical or chemical magnitudes, but it includes any device which can provide useful information, such as user's mobile devices acting as logical devices. SEAC is supported in:

- Physical-Digital world data synchronisation and exchange
- Multi-domain, context-sensitive seamless access control for fast decision making

SEAC establishes that access control management has to be a flexible and smart procedure that coherently takes advantage of all potentially useful information sources to increase certainty in decision making, pondering the information sources with respect to their associated (configured) weight for identification and usability of the system. Under the SEAC concept, already deployed information sources have to be analysed and included into the decision making system and additionally required identification mechanisms are selected giving priority to usability. The aim is to increase the number of information sources that will not degrade the system's usability instead of deploying highly secure mechanisms that affect severely fluency. The main aim is to keep the same security level but higher fluency and usability.

3.2 ACIO Framework Principles

ACIO principles are set by the SEAC concept. There are four main principles identified in the ACIO framework:

- The access control management should be smart; it should learn over time, anticipating risk situations and detect anomalous behaviours or patterns to act accordingly. Note that self-learning and proactive detections have always been key aspects in security management, especially in access control.
- Access Control management must maximize the incorporation of relevant information sources, multiplying the information sources to enrich decision making and the security level of the system, without degrading usability. This relates to a scalable and interoperable module able to integrate any potential information source (sensor in a general sense).
- The incorporation of new information sources must be coherent, weighting in each case its contribution to decision about access granting against complexity of its integration to the system and to the overall processing. Just adding new sources does not necessarily help the decision making process and adds extra complexity; therefore, the relevance of any aggregated source must be previously analysed and given a certain weight on each implied access control policy.
- Decisions for Access granting must be balanced between security and usability, proposing or executing additional actions of control, rather than access decisions which will degrade fluency. The capabilities of new technologies should be seized to complement the access control.

The previous principles promote the development of a flexible and smart access control management system that copes with the current requirements of organizations.

3.3 ACIO Proposed Architecture

The proposed ACIO architecture searches for a throughput improvement of 2 or 3 times in respect to traditional systems in terms of number of users identified, without major infrastructure investments, providing full traceability of people and assets while preserving and evolving the already implemented legacy systems in case they exist.

The systems developed under the ACIO framework, with a user centred access control management, provide:

- Automated data acquisition
- Accelerated (multi-domain) cross system and cross platform data exchange
- Fast and easy (multi-factor) access control policy delegation with traceability
- Big data context-related segmentation for fast analysis
- Seamless physical-digital world cloud-based data exchange management
- Digital evidence management

Figure 1 provides a scheme of the generic architecture for an ACIO framework Access Control management system. The main component of the architecture is the Authorization System, this central neuralgic element is responsible for the access control decision-making in real time through the collection of information from the different worlds (physical and logical), as well as for the communication bus implementation among the involved entities and modules.

Following the principles of the ACIO framework, the authorization system gathers and processes relevant information from multiple sources in order to provide access granting decisions, generate alerts and trigger concrete control (proactive and reactive) actions. The architecture encompasses the following information sources (see Fig. 1):

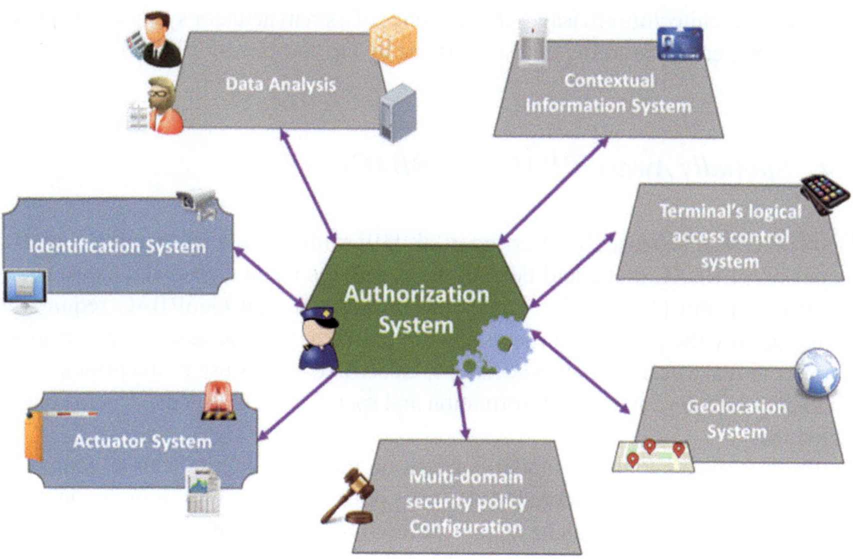

Fig. 1 ACIO framework proposed architecture

- *Identification System*: Composed by physical identification mechanisms, such as biometric sensors, vehicle identification devices, container and plate number reading mechanisms, etc. These are just examples of common identification devices employed in the port of Valencia (used as testbed), but any other ones are also valid.
- *Contextual Information System*: it is responsible for the collection, processing and storing of all contextual information generated, and of ensuring the interoperability and sharing of all information available in order to avoid duplicates or inaccuracies. Note that context knowledge is interesting not only in real time, but also offline to better analyse past events and anticipate future actions.
- *Logical access control system*: it involves all logical identification systems that provide digital means of identification: credentials, tokens, digital certificates, etc. are included under this category.
- *Multi-domain security policy Configuration*: this system processes the security policy for all interconnected domains, providing information about how the policy must be applied in each case.
- *Data analysis:* This information source includes all systems capable of the following features: threat detection, pattern matching and statistical correlation. Obviously the required data to be processed may come from other modules (e.g. Contextual Information System).
- *Geolocation System (logical)*: it involves systems that provide information about positioning that enable people or objects tracking between different domains and locations. This is especially interesting for GeoRBAC mechanisms, as will be explained in the next subsection.

Finally, the architecture envisages the existence of a smart actuator system that applies the resulting access granting decisions made by the authorization system.

3.4 Spatially Aware RBAC (GeoRBAC)

GeoRBAC was proposed by Bertino et al. [19] with the aim of creating a spatial extension of RBAC that could be compliant with the Open Geospatial Consortium geometric model [1, 19]. The contextual information that GeoRBAC requires is minimal, only the position of the user will be used as complementary information. In order to make access control decisions, GeoRBAC makes use of the principles of set theory with the contextual information and spatial roles to execute access control decisions.

The reference geometric model is based on the OGC simple feature geometric model, where any object is composed by one or more points, lines or polygon types and topological relations can be applied to them, such as specifying different granularities to defined objects. Independently of the technology of the tracking devices, the model can assign logical positions to each user based on their real positions through different mapping functions. For example, a device providing real-time

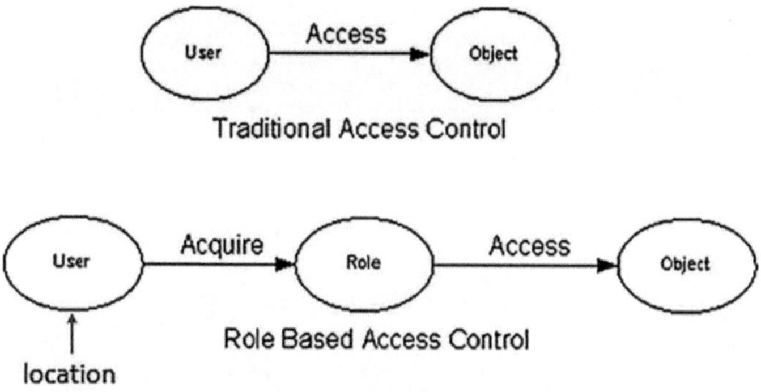

Fig. 2 Generic (simplified) RBAC model including location

three-dimensional coordinates can be mapped to a specific room and granularity, identified by a logical label in the position model.

In order to maintain compatibility with traditional RBAC models, GeoRBAC defines spatial roles as role extensions, being each spatial role a representation of a geographically bounded organizational entity, and roles that don't consider spatial boundaries will be given a role extent that considers the full reference space (Fig. 2).

Spatial roles as role extensions not only gives backward compatibility but also ensures that GeoRBAC can be extended with other models that enhance the context awareness in the form of role extensions. That is the case of ProxRBAC, which introduces proximity constrains in the access control decisions or TRBAC, which provides the temporal capability that GeoRBAC lacks, and is essential in many applications that require spatio-temporal constraints [7, 8].

GeoRBAC supports role schema and role instance hierarchies that allows inheritance of permissions, user assignments and activation between roles; policy administration is not part of GeoRBAC, as well as policy integration or multiple object attributes, although some of these were addressed in later proposals.

4 ACIO Use Case

A specific solution defined under the framework to be implemented in a real world scenario has been used to demonstrate the ACIO framework's performance and benefits.

This use case refers to the dependencies of the Port of Valencia. Valencia seaport is an ideal scenario as CI and it encompasses highly complex control requirements both at physical and logical access control level. The port use case provides interaction among a great variety of actors and heterogeneous elements (e.g. containers, machinery, trucks of different operators, etc.). Additionally, there is a high dependency of the port's competiveness regarding fluidity of the ongoing operations, and the obligation

Fig. 3 Valencia port overview

to comply with different rules and national and international legislation all of which set the success indicators for the results. The port premises allocate five container terminals (see Fig. 3) with more than 4000 truck movements and 10,000 containers per day. The port operations are management through a PCS (Port Community System) in the cloud, which enables and conducts information exchange among all port actors (port authority, customs, transport companies, etc.) and interacts in a federated way with their systems (see Fig. 4).

To set the use case, the first step was to analyse the existing access control to detect deficiencies, strengths and potential improvement points, as well as gathering the port authorities' needs and the facts and figures regarding operations. As result of these previous evaluations, a number of deficiencies were detected and a solution based on the ACIO framework was analysed and proposed to tackle the previous weaknesses and are listed in Table 2. As can be seen, the use case will enable the evolution from the currently static and inefficient AC management of the port, with consequently low security, to a user centred AC management providing a high security level and high usability which complies with seaports requirements.

The proof of concept for the ACIO framework has been carried out in two phases. The first phase has been done under laboratory conditions and has focused in assuring correct component functioning and integration, as well as solutions scalability through extrapolation of results registered by processing power requirements.

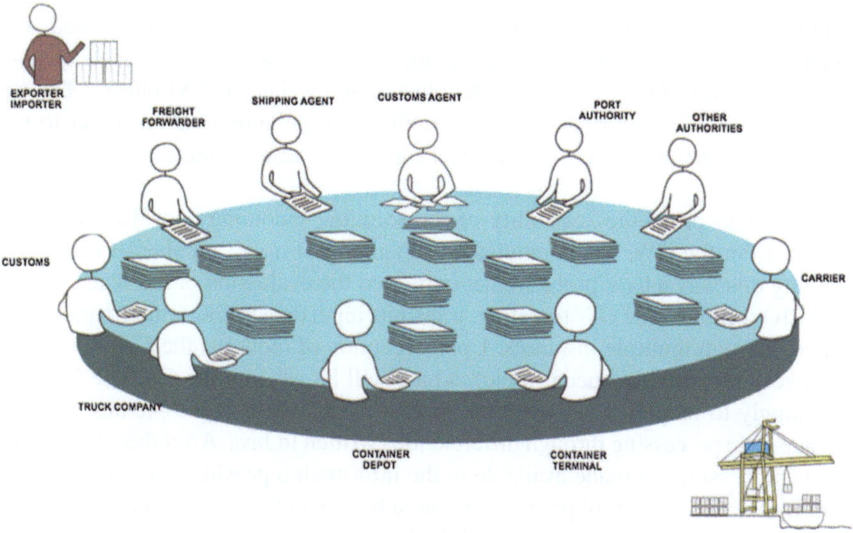

Fig. 4 Port problematic (information flow) solved via PCS

Table 2 ACIO framework based AC management improvements

ACIO framework approach in the valencia port AC management system		
AC process	Current AC management system	ACIO improvements
Registration	Repetitive and heavy	Agile process only done once
Identification	Identification of all users is currently not being done as it would affect fluency and does not support mobility	Multifactor, unified and mobile
Authentication	Interacting with the PCS requires multiple checkpoints with their consequent credentials	Cross-domain single-sign-on
Authorization	Current authorization process is slow and does not consider any additional information sources (context)	Transparent decision making and dynamic policy based in context
Accounting	Manual and incomplete	Complete and automated traceability and awareness
Auditing	Poor compliance evidence	Automated governance and compliance evidence

For these demonstrations, a test environment shown in Fig. 3 has been used. The Valencia's specific AC management solution encompasses the following components: a gateway (GW), a Kafka (CK) cluster, and a Samza (CS) cluster. The CS provides data to an information provider, allowing the authorization server to use this information in the process of physical and logical access control in an integrated way.

In order to prove the scalability of the solution, incoming data was generated directly from the GW, which emulated exchanging of a vast number of messages; allowing this way a base problem that permitted the evaluation of the scalability of the ACIO solution. The GW therefore, transmits input data to CK, which deployment is done through multiple instances. Upon reception of requests, the CK ordains the messages on a specific queue (topic), which will be used by the CS. The CS will, accordingly to its policy subscription, extract the message of a topic and perform cascade style processing through different jobs written in Java. After this, the outputs of this processing are made available to the information provider; that will provide them for the access control process, executed by the authorization server.

The current measured operations of the Valencia Port, in sensing terminals, state a flow of about 8,000 messages/second. Validation carried out deploying the cluster in general purpose systems, allowed the processing of 21,456 messages per second, four times the current activity, using only 35% performance capacity. Taking into account the transparent scalability of the system, the ACIO solution has not only been validated to its deployment in Valencia's port, but in scenarios of equivalent growth and evolution. Worth mentioning in this sense, that the port of Valencia is currently among the ten busiest in Europe in number of containers.

The final demonstration of the viability of the solutions developed under the ACIO framework encompasses the deployment of a sample ACIO solution in several checkpoint and allocations of the port premises to evaluate the systems performance and behaviour under real conditions. The use case will be deployed in 4 locations: (1) North access gate to the port, (2) Container terminal, (3) Cruise terminal and (4) Infoport building (one of the companies involved in the ACIO framework specification which is a main IT solution provider to port actors), these locations are shown in Fig. 5.

Regarding the logical access control, the pilot shows the integration of the solution with Sintraport, an IPV application for inland transport management, interoperating with the Port Communication System (PCS) in the Port of Valencia:

- Transportation and logistics companies which use Sintraport can send to the PCS all the information related to the vehicle, the cargo, the driver, the load/unload points and even the expected time schedules.
- All these data can be sent to the authorization server to manage the policies.
- Companies introducing information for Sintraport will be marked as "secure" if they keep ACIO security policies, as an added value.

Unfortunately, the interaction of the ACIO solution components with the existing ones is unfeasible due to legislation constrains.

Fig. 5 Locations of the ACIO solution deployment in the use case

ACIO solution will be deployed therefore as a partial isolated system. Currently, accessing the port requires showing the Valencia port authorization card, a card allocated by the Port Authority of Valencia to the port employees.

The main parameters that will be used to evaluate the performance of the ACIO solution will be:

- Delay: comparison of time needed by users to surpass checkpoints. This allows us to measure system fluency.
- Identification: Number of users identified. This allows us to measure throughput.

The ACIO components selected to be deployed will be:

1. Sensors deployed:
   (a) In-motion identification enabled by portable smart cameras.
   (b) LPR (License Plate Recognition) camera to identify car plates.
   (c) RFID reader to detect RFID tags that will allocated in the vehicle to identify the container and the car plate.
   (d) Smartphone with GPS activated and tracking mobile application installed.
2. Network and power enablers:
   (a) 4G wireless router and an Ethernet switch for the Internet access.
   (b) Power supply element (generating set).
3. Actuators:
   (a) Traffic lights and virtual barriers (microwave, photocells, etc.)

Table 3 ACIO expected improvements respect to the current use case AC managements system

Location	User type	Identification time (AVG)	
		Current AC	ACIO
Main gate	Transporters	30–40"	8–10"
	Usual workers	8–10"	3–10"
	Outsources and visitors	30–40"	8–10"
	Passengers	30–40"	15–40"
Terminal	Transporters	25"	8–10"
Buildings	Usual workers	5"	3–5"
	Outsources and visitors	40–50"	5"

Table 3 shows the expected results for the ACIO solution to be deployed against the current AC system.

4.1 Use Case Defined Roles:

(a) *Visitor*: This role represents a person who doesn't visit the port frequently. Three possible sub-roles are considered:

  (i) *First time visitor*: A person who enters in the port for the first time.
  (ii) *External company*: An employee of an external company who needs access to the port area to visit a port company.
  (iii) *Passenger*: A person who is going on a cruise and has to access to the port cruise terminal.

(b) *Habitual visitor*: This role represents a person who has to access to the port area frequently. For example, the personnel of the Maintenance service: A person in charge of carry out maintenance tasks within port area.
(c) *Employee*: This role represents a person who, due to his job, has to access to the port area every day.
(d) *Security authority*: A person who is part of a port security authority.
(e) *Port employee*: A person who works in a company located within port area.
(f) *Truck carrier*: This role represents a truck driver who must access to the port to load or unload cargo in the terminal:

  (i) *Container*: The cargo is a container:
  (ii) *General cargo.*

All roles, except employee role, have to register their visit before access to the port. Depending on the destination, users will be registered through a different Identity Provider.

4.2 *Use Case Decision Making:*

The ACIO solution will make different decisions related to AC depending on the information received by the different systems deployed in each of the locations, the location type, the security politics, etc. For the demonstrator however, the only parameters that are going to be taken into account for AC decisions are the information of the deployed AC mechanisms and the location of the access control decision point.

For the scope of the use case, and due to the impossibility of taking control of the current actuators deployed in the port, the decisions made by the system will be presented through several visual interfaces.

In this same line, the possible decisions to be taken by the system will be limited for the sake of demonstration fluidity. The outputs of the authentication server regarding the AC decisions will therefore be limited to 5 possible states depending on the information received:

(g) *Access granted*: This means that the person requesting access to the port is known and authorized or unknown but trusted. The system will keep a record of unknown people that will enable identification if a security event is detected.
(h) *Manual identification required*: The system detects suspicious behavior, untrusted user or other anomalies, and request the Port authority to manually identify the person, this person will be forced to deviate to a security zone where it will be requested to identify itself.
(i) *Grant access but track activity (GEO-RBAC)*: The system detects an abnormal event. For example, a known authorized user with an untrusted/unexpected behavior and enables the GEO-RBAC tracking of the requester.
(j) *Grant access and send alarm*: The system will record access requests that present an anomaly but don't seem suspicious (e.g. authorized and expected user in an authorized but not expected vehicle).
(k) *Deny Access*: Finally, blacklisted people will not be permitted to access the port.

5 Conclusion

The validation tests at laboratory level have demonstrated that a solution based in the ACIO framework enable an integrated and holistic access control management in high demanding environments, assuring the system feasibility, even when using general purpose equipment.

On the other side, the ACIO approach permits the optimization of access control efforts and usability, by proposing the incorporation of additional information sources that contribute to perform consistent decision making for access granting in highly sensor available environments.

The solution deployed has demonstrated that a solution based in the ACIO framework is capable to cope with the current requirements of the most demanding business processes. The authorization server gathers information of the deployed information sources (sensors) and reacts to different situations according to the defined actuation protocol. The system is supported by a geo-tracking system, enabled by a mobile application that serves to add further control capabilities, avoiding fluency operations interruptions or long delays when anomalous situations occur, such as authorized user driving an unknown or unexpected vehicle or vice versa.

However, there is further research in the access control management to do. Although the Physical-Digital world data synchronization & exchange and Multi-domain, context-sensitive seamless access control for fast decision making features that the ACIO framework is capable of providing are significant advancements with respect to the existing management systems for access control in organizations, the evolving nature of the IoT concept along with its challenges and the sensing capabilities that are being incorporated to the "sensing enterprise" open a unknown path fraught with uncertainty and risks and of course possibilities. This is because these systems present a dichotomy between security policies (e.g. confidentiality, integrity) and commercial policies (e.g. usability). Therefore, traditional approaches are rarely capable of providing the required security when business processes impose flexibility, fluidity and collaboration with other entities.

References

1. Skinner, G.D.: Cyber security management of access controls in digital ecosystems and distributed environments. In: 6th International Conference on Information Technology and Applications (ICITA 2009), November (2009)
2. The OGC Sensor Web Enablement (SWE), Open Geospatial Consortium (OGC). http://www.opengeospatial.org/ogc/markets-technologies/swe/ (2013)
3. Sensor Observation Service (SOS), Open Geospatial Consortium (OGC). http://www.opengeospatial.org/standards/sos (2013)
4. Giménez, P., Molina, B., Palau, C.E., Esteve, M.: Sensor web simulation and testing for the IoT. In: IEEE International conference on Systems, Man, and Cybernetics (IEEE SMC 2013), Manchester, October (2013)
5. Karp, A.H.: Authorization-based access control for the services oriented architecture. In: 4th International Conference on Creating, Connecting, and Collaborating through Computing (C5), January (2006)
6. eXtensible Access Control Markup Language Version 3.0, OASIS XACML v. 3.0, August (2010)
7. OpenID Connect Core 1.0. http://openid.net/specs/openid-connect-core-1_0.html
8. Internet Engineering Task Force (IETF). Request for Comments: 6749. The OAuth 2.0 Authorization Framework. http://tools.ietf.org/html/rfc6749
9. OASIS Standard. Security Assertion Markup Language (SAML) V2.0. http://docs.oasis-open.org/security/saml/v2.0/ (2005)
10. Ferraiolo, D.F., Richard Kuhn, D.: Role-based access controls. In: 15th National Computer Security Conference, pp. 554–563 (1992)
11. Yuan, E., Tong, J.: Attributed based access control (ABAC) for web services. In: Proceedings of the IEEE Conference on Web Services (ICWS 2005), pp. 561–569 (2005)
12. Shen, H.: A semantic aware attribute-based access control model for web services. In: Lecture Notes in Computer Science, vol. 5574, pp. 693–703 (2009)

13. Shen, H.: A semantic context-based access control model for pervasive computing environments. In: Proceedings Advances in Intelligent and Soft Computing, vol. 168, pp. 135–140 (2012)
14. Gunter, C.A., Liebovitz, D.M., Malin, B.: Experience-based access management: a life-cycle framework for identity and access management systems. **9**(5), 48–55 (2011)
15. Liu, D., Li, N., Wang, X., Jean Camp, L.: Security risk management using incentives. In: Proceeding IEEE Security and Privacy, November, pp. 20–28 (2011)
16. Piccione, S., Gusmeroli, S., Rotondi, D.: A capability-based security approach to manage access control in the internet of things. Math. Comput. Model. **58**, 1189–1205 (2013)
17. Fortino, G., Trunfio, P.: Internet of Things Based on Smart Objects, Technology, Middleware and Applications. Springer, Berlin (2014)
18. Nguyena, K.T., Laurentb, M., Oualha, N.: Survey on secure communication protocols for the Internet of Things. Elsevier Ad Hoc Netw. **32**, 17–31 (2015)
19. Bertino, E., Catania, B., Damiani, M.L., Perlasca, P.: GEO-RBAC: a spatially aware RBAC. In: Proceedings of the Tenth ACM Symposium on Access Control Models and Technologies (SACMAT '05), pp. 29–37. ACM, New York. http://dx.doi.org/10.1145/1063979.1063985 (2005)
20. Kirkpatrick, M.S., Damiani, M.L., Bertino, E.: Prox-RBAC: a proximity-based spatially aware RBAC. In: Proceedings of the 19th ACM SIGSPATIAL International Conference on Advances in Geographic Information Systems, GIS '11, pp. 339–348. ACM, New York (2011)
21. Aich, S., Sural, S., Majumdar, A.K.: STARBAC: Spatio-temporal role based access control. In: Proceedings of the 2007 OTM Confederated International Conference on On the Move to Meaningful Inter net Systems: CoopIS, DOA, ODBASE, GADA, and IS - Volume Part II, OTM'07, pp. 1567–1582. Springer, Berlin (2007)
22. Sandhu, R.S., Samarati, P.: Access control: principle and practice. IEEE Commun. Mag. **32**(9), 40–48 (1994)
23. Crampton, J., Morisset, C.: Towards a Generic Formal Framework for Access Control. CoRR vol. arXiv:abs/1204.2342 (2012)
24. Rotondi, D., Seccia, C., Piccione, S.: Access Control and IoT: Capability Based Authorization Access Control System, November (2011)
25. Skinner, G.D.: Cyber security management of access controls in digital ecosystems and distributed environments. In: 6th International Conference on Information Technology and Applications (ICITA 2009), November (2009)

The Application of Telematics and Smart Devices in Emergencies

Marco Manso, Barbara Guerra, Cosmin Carjan, Evangelos Sdongos, Anastasia Bolovinou, Angelos Amditis and David Donaldson

Abstract The adoption of smart environments is becoming more and more important in many application scenarios such as healthcare, asset management and environmental monitoring. In emergency services, there are also very attractive use cases on next generation emergency services, presenting challenges that must be addressed in order to satisfy the requirements of both citizens and emergency service professionals. This chapter shares two interesting use cases depicting emergencies where the use of telematics and smart devices do enable improved emergency situational awareness for citizens and emergency services. Leveraging on the example of the eCall initiative, a thorough explanation is given on how next generation emergency services may further explore the promising new smart technologies to improve the communication dynamics with citizens and to attain increased effectiveness and performance in their daily mission to protect and safeguard lives.

M. Manso (✉) · B. Guerra
RINICOM, Lancaster, UK
e-mail: marco@rinicom.com

B. Guerra
e-mail: barbara@rinicom.com

C. Carjan
Teamnet International, Bucharest, Romania
e-mail: cosmin.carjan@teamnet.ro

E. Sdongos · A. Bolovinou · A. Amditis
Institute of Communication and Computer Systems, Athens, Greece
e-mail: esdongos@iccs.gr

A. Bolovinou
e-mail: abolov@iccs.gr

A. Amditis
e-mail: a.amditis@iccs.gr

D. Donaldson
West Yorkshire for Innovation Team, Police Crime Commissioner for West Yorkshire, Wakefield, UK
e-mail: david.donaldson@westyorkshire.pcc.pnn.gov.uk

R. Gravina et al. (eds.), *Integration, Interconnection, and Interoperability of IoT Systems*, Internet of Things, DOI 10.1007/978-3-319-61300-0_9

1 Introduction

The origins of research in the Internet of Things (IoT) field are closely related to the many challenges experienced with embedded and cyber-physical systems, pervasive computing, semantics, integration of network technologies and sensors. It was indeed the need for convergence in all those domains that the IoT concept aims to capture and solve, acting as an umbrella term with the purpose to unite the underlying technologies and their related vision [1].

The IoT concept refers to the extension of the Internet paradigm to the world of objects and places, each able to communicate their own data and access aggregated information from other objects and places. Building blocks of the future IoT, smart objects cooperate to fulfil specific goals [2], displaying the ability to sense environmental parameters that is becoming increasingly important in diverse application scenarios, a trend ambitioning to inspire smart environments and capture, in a pervasive way, useful information from the real world, contributing to assert the IoT concept. In this way, the new machine-to-machine paradigm aims to improve everyday life: healthcare (smart health), cities management (smart city) and energy saving (smart energy) are typical scenarios for the use of such applicable technologies.

Albeit the mobile and fixed IP networks have increasingly penetrated over the last years in the consumers' daily communication needs, it has not been the case for emergency services. Public Safety Answering Points (PSAP) operators, emergency services' professionals and citizens in distress continue to experience hurdles in their bidirectional communication dynamics, namely when using next generation (NG) IP-based technologies and networks and upholding non-traditional information exchange. These hurdles are in essence determined by the lack of network and application interoperability across the emergency communication chain. Further, this *disconnected* reality is heightened with the proliferation of different models for handling/dispatching emergency calls throughout Europe [3], often based on proprietary implementations.

To address the interoperability barrier, the NG emergency services concept brings forth a standardised IP-based network for emergency services capable of providing dynamic call routing of multimedia communications and advanced call processing functions. Recent publications depicting the status of PSAPs in Europe [4] yield that only minimum usage of capable Internet-based technologies (e.g., IP) is currently realised, whereas the deployment of services other than the traditional voice calls, encapsulated in the Total Conversation paradigm (the rich combination of voice call, imagery, video footage and real-time texting), is sporadically implemented.

As society begins to evolve towards next generation emergency services, many emergency-related applications for smartphones have emerged worldwide. Where-AREU by AREU,[1] SafeBeep by Vodafone[2] or ICE by Matrix Mobile Applications[3] are examples of applications that provide assistance in emergencies, crises or

[1] https://where.areu.lombardia.it.

[2] https://www.safebeep.gr/#about.

[3] http://www.emergencyapplication.com/.

disasters to affected citizens, including those experiencing disability, impairment or special needs, through the exploitation of ubiquitous connectivity that enable enhanced situational awareness to those offering aid.

Notwithstanding, numerous functional and operational issues remain yet to be resolved to the benefit of both emergency services and citizens. And a particular important issue refers to the application of telematics and smart devices in emergencies and its adequate adoption by next generation emergency services.

Indeed, among the potential applications of smart societies, the use of telematics and smart sensors and devices in emergencies is deemed particularly important, due to its capability to empower the automated early detection of emergencies, including in situations where human intervention might not be possible or is impaired, as well as an enhanced emergency situational awareness, benefitting from more accurate location data and the real-time exchange of relevant information. Overall, the adoption of the IoT paradigm by next generation emergency services would represent a new, innovative and valuable link between citizens and emergency services that would greatly improve the quality of emergency response worldwide.

The NEXt generation Emergency Services (NEXES[4]) Research and Innovation Action (RIA) [5] is currently tasked by the European Union to design, implement and validate a reference implementation of next generation emergency services, benefitting from the promising integration of IP-based communication technologies and interoperability into emergency services. In this context, NEXES is bound to explore the benefits, shortcomings and challenges involving the connection of smart sensors and devices to emergency services. It is expected that, once smart sensors' data is integrated with the emergency call and automatically relayed to emergency services, this new added-value feature is able to improve these services' capability to assess the emergency and respond swiftly and more effectively, protecting and saving lives.

2 On the Path Towards Next Generation Emergency Systems

According to the Coordination Group on Access to Location Information by Emergency Services, each year in the European Union several millions of citizens dial an emergency call number to access emergency services [6]. The fundamentals on emergency services highlight that communication between citizens and emergency services rely mainly on voice calls and that only partial accessibility support for citizens experiencing disability or impairment is provided through SMS and fax. Caller location is carried out by all emergency services but the accuracy and precision of the caller's location still remains the biggest hurdle of current emergency services,

[4]The NEXES Action has received funding from the European Union's Horizon 2020 Programme for research, technological development and demonstration under grant agreement no. 653337. More information available at http://nexes.eu.

leading to considerable time loss: it is estimated that emergency services are not able to dispatch a rescue team for approximately 2.5 million calls due to the absence of sufficient location information [7]. In addition, it is widely recognised the emergency services' ambition to adopt new technologies, interoperability and telematics to improve their efficiency, effectiveness and performance and build a more secure society.

With a century old research history, the sociological study of emergencies is aware that information and communication technologies are able to expand the reach of disaster sociology [8] and is now focusing on the promising benefits of integrating IP-based communication technologies and interoperability with traditional emergency services, following the strongest market pull-drive ever. Notably, this integration foresees a true leap forward in the way PSAPs, in what respects emergency calls' answering and emergency information exchange, are able to fulfil their obligations towards citizens in need through a plethora of (over IP) communication channels that brings about richer communication (voice, text, imagery, video and data) and insightful multimedia information on numerous platforms (including social media and mobile applications).

The NEXES RIA aims to research, test and validate the integration of IP-based communication technologies and interoperability into the next generation emergency services (NGES), so that they attain increased effectiveness and performance. NEXES proposes the introduction of a set of new, cutting edge Internet-based technologies to dramatically improve the way citizens and emergency services exchange information and develop accurate awareness (who, what and where) on emergencies.

NEXES identifies as promising key enablers for NGES:

- **Total Conversation (TC)** – A standardised concept allowing an audiovisual conversation with bidirectional full-duplex real-time transfer of real-time text (RTT), video and voice. The use of video can, for example, allow PSAPs to better assess the caller's condition, the magnitude of a fire or an accident. Citizens would use real-time text when in special circumstances, such as a noisy environment, in case of network overload caused by a large-scale disaster, large public unrest or severe power outage or even in life-threatening situations, when text is safer than voice to reach emergency services. TC is also especially well-suited for citizens experiencing disabilities or impairments: deaf citizens would communicate via real-time text or use video stream with sign language; hearing-impaired users could benefit from real-time text or video stream (lip reading) to communicate with the emergency call operator; and deaf-blind users would rely on real-time text to communicate via an appropriate input/output device.
- **Device data** – To respond to an emergency request, current emergency systems rely on information provided by the caller and by the telecommunications operator, mainly consisting in caller identification, name and location. However, modern User Equipment (UE), such as smartphones and tablets, are able to send additional data available in the device (vcard, medical history, personal profile and Global Navigation Satellite System or GNSS information) that can be used to better prepare the emergency response. Although the incorporation of UE data is now being

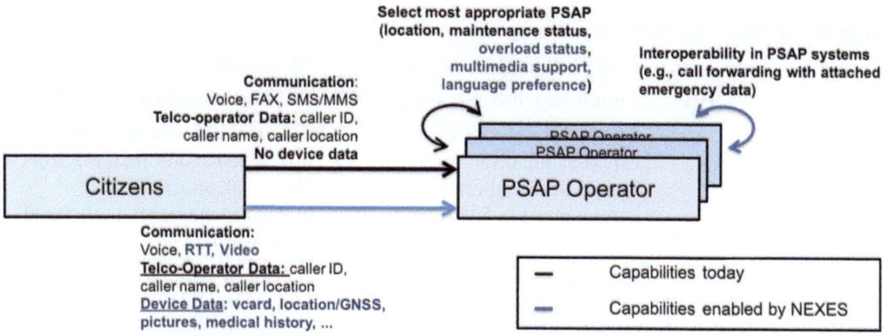

Fig. 1 Emergency services today and benefits envisioned in NEXES

explored by different emergency services, additional developments on emergency calls data are expected following the work being performed by the Internet Engineering Task Force (IETF) [9].

- **Location-aware devices** – Accurate, precise and reliable location information on emergencies is still a subject of major concern to emergency services, especially in incidents where callers are not able to report it or simply do not know it. Most UE today have location-aware capabilities, either via embedded GNSS, WiFi Access Point (AP)[5] or cell ID/sector. With the caller's consent, this information may be automatically incorporated in the emergency call, as part of the device data transmitted to PSAPs, and greatly improve the emergency location information.

Figure 1 and Table 1 present the enhanced communication between citizens and emergency services (PSAP operators in particular), today and as envisioned by NEXES.

NGES brings enhanced communication and end-to-end connectivity between citizens and emergency services, using TC calls, and provide emergency services with more data, such as device location, telematics and health data, to improve the level of situational awareness and enable coordinated emergency management. An overview of NEXES is presented in Fig. 2. Four main vertical domains are presented, comprising the citizens (call origination), the public communications infrastructure providers (e.g., telecommunication operators and Internet Service Providers), the emergency services' IP network and the PSAPs (call answering). In this chapter, the focus is placed upon the aspects related with call origination and the information exchange with PSAP operators.

In detail, Fig. 2 illustrates in green the components that today are part of most emergency services across Europe, in orange the components that have been implemented in just a few emergency services (e.g., SMS) and in red the components that, although existing in PSAP systems, are not yet operational in emergency services (e.g., eCall). Figure 2 also presents the new components to be developed within

[5]The low range of WiFi AP (typically 100 m) is an acceptable location accuracy for most emergencies.

Table 1 NEXES improvements to emergency services

Parties	Today's emergency services	NEXES emergency services
Citizens <–> PSAPs	Circuit switch-based voice calls	IP-based broadband networks
	Unequal access of citizens experiencing disabilities and special needs (alternative communication SMS, Fax)	TC capabilities (audio, real-time text, video and/or images) Universal, accessible and inclusive emergency tools for all citizens
	Limited call location capability	Enhanced location capability (device location)
	No UE Data	Enhanced data exchange (device data, vCard, medical history)
	Limited multilingualism capability (languages of neighbouring countries)	Enhanced multilingualism support (nationality shared by citizen profile)

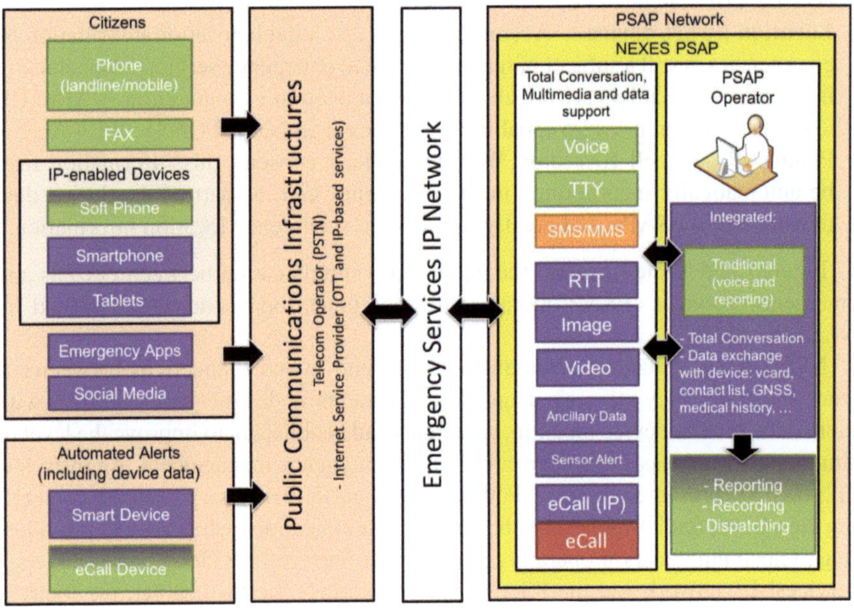

Fig. 2 NEXES overview (adapted from [10])

the NEXES RIA in purple and the existing components that require adaptation and extension work to fully exploit the new functions brought by NEXES in gradient green-purple.

Identified as "Automated Alerts" and counted amongst the most promising key enablers of next generation emergency services, telematics and smart sensors and devices exhibit significant relevance in building emergency situational awareness and empowering higher performance in emergency response.

NEXES's concept of smart automated emergency services is not a competitor to or a new implementation of existing emergency communication channels, but rather a supporting technology providing added value in terms of:

- **Interoperability among emergency services** – In NEXES, a web of IoT platforms will be created to provide the necessary infrastructure for fostering interoperability among various emergency services, such as law enforcement, firefighters or medical assistance, in the local, regional, national and cross-border levels. Some of these services already utilise IoT platforms for their applications and alerts, while others cannot currently be reached with means other than standard voice communications;
- **Communication between citizens and emergency services** – Within NEXES, citizens and emergency services are expected to gain access to the applications developed, thus allowing mutual bidirectional communication. Emergency agencies shall be able to bidirectionally reach citizens via IP-based applications that exploit IoT systems, smart devices and smart environments, either on demand or on automated modes.

Amongst the panoply of smart technologies to be explored in NEXES, the following three are considered to be aligned with the purpose of this chapter:

- **eCall** – The eCall concept is currently deployed based on circuit switched emergency calls, specifically for 2G and 3G networks using an in-band modem [11]. As next generation emergency services evolve towards a 4G communications framework, the eCall concept will be refined and adapted in NEXES using Session Initiation Protocol (SIP) and Minimum Set of Data (MSD) extensions, hence enabling the benefits of resource-efficient packet-based systems.
- **eHealth Devices** – NEXES will explore the enormous potential of eHealth devices and Apps in its interconnection to next generation emergency services. Not only individuals' health data (vital signs, heart rate, temperature) may be relayed to emergency services for enhanced emergency situational awareness but also automated emergency calls can be initiated in case a severe health trigger is detected. Similarly to eCall, NEXES will explore SIP emergency call and MSD extensions in the connection to smart eHealth devices.
- **Smart Environmental Sensors** – The growth and adoption of smart environmental sensors and devices, such as fire, gas and seismic sensors, will result in (smart) houses, factories, buildings, communities and cities that are environmental-aware of their status and anomalies. NEXES will explore the automated connection of those environmental sensors to emergency services: integrated with the emergency call, environmental sensors' data (temperature, gas and movement intelligent readings) are automatically sent to emergency services, improving their capability to assess the emergency.

Automated and data-enabled emergency calls will be explored in NEXES, relying on sensor data retrieved from vehicles, buildings and even individuals. Within NEXES, sensor data will be decoupled from the sensing devices (related to the type of measured data) and the respective events/alerts they produce. SIP capabilities will be analysed to handle sensor data and define a framework for automated sensor-based emergency calls. Data structures and mechanisms will follow work in IETF's Next-Generation Pan-European eCall document [12], which includes: location, call-related information (device type, sensor type, service providers, subscriber details, personal information, contact information) and caller information (medical information and *in case of emergency* or ICE contact data). Following this path and reviewing other relevant standards and initiatives, NEXES will compile a fusion of additional data elements, conveying innovative proposal for additional data structures capable of addressing automated and data-enabled emergency calls.

3 Telematics and Smart Devices for NGES

The developments towards the wide proliferation and broad adoption of sensors, telematics technologies and the IoT preclude a future highly automated and environmental-aware, namely in what respects to the automatic detection and to the enhanced situational awareness of emergency incidents that have the potential to be critical elements for fast and effective emergency response, in the protection of citizens and the safeguard of lives.

Hence, it is expected that NGES will support automated emergency calls initiated by smart devices that may rely on sensors installed in vehicles (e.g., eCall systems) or in residences and facility buildings (e.g., seismic sensors, gas sensors and fire sensors) and in wearable health devices (e.g., heart rate monitor, fall detection). Aside from being capable of automatically initiating an emergency call, supporting voice and video when in the presence of the victim, smart devices will exhibit the ability to provide relevant data on the caller and the event to be interpreted by PSAP operators. Automated and data-enabled emergency calls are two examples of the major changes in paradigm in the traditional way of handling emergency calls worldwide.

To adequately understand the impact of telematics and smart sensors in emergency services and better explain NEXES's approach to the integration of new automated and data-enabled emergency calls, the NEXES RIA started by defining reference use cases where these new functions would undoubtedly represent an added-value for society [10]. Leveraging on the extensive experience of its wide user community, NEXES's reference use cases, presented next, thus guarantee a common vision to steer the effort of incorporating IoT innovation into the next generation emergency services.

3.1 Reference Emergency Use Cases Using Telematics and Smart Devices

Use Case 1: Traffic Collision Involving Citizen with Impaired Speech and a Vehicle with eCall

Person 1 has recently undergone dental surgery and is suffering temporarily from limited speech. Driving along a remote rural road, the vehicle (vehicle 1) collides with another car (vehicle 2). The driver (Person 2) of vehicle 2 is rendered unconscious in the collision. Person 1's vehicle sustains extensive damage to the front offside and, although conscious, Person 1 is trapped in the seat, with a painful right leg.

Person 1 immediately initiates a 112 call through the NEXES App on their smartphone. Due to the recent surgery, Person 1 is unable to clearly communicate the emergency to the emergency call operator, who issues a request for an instant messaging conversation or a video call. Person 1 accepts the request for a video call and, as the connection is established, Person 1 shows the emergency call operator footage of the accident and of vehicle 2. The emergency call operator is able to observe that Person 1 is trapped and that a second victim seems to be unconscious in vehicle 2. Person 1 consents to share their mobile device's location information with emergency services and the call operator is thus able to retrieve the caller's exact location.

Simultaneously, an eCall from vehicle 2 is automatically initiated to the emergency services after the collision. The emergency call operator calls back to confirm the emergency but receives no answer. However, because the location of the vehicle provided through the eCall matches the location of Person 1, provided through the NEXES App, the emergency call-handling system concludes that both calls relate to the same incident. Having gathered all relevant data, the emergency system links the calls and passes the information to the appropriate emergency response organisations that dispatch their First Responders (FRs) to the collision's site. Firefighting, medical and law enforcement FRs en-route now have the benefit of sharing the information gathered by the call-handling centre via the NEXES IP-enabled eCall and video link, thus maintaining an enhanced emergency situational awareness, enabling improved coordination amongst emergency services and providing a more efficient and effective emergency response.

Use Case Highlights:
This scenario emphasises numerous facets of NEXES's capabilities. The Total Conversation function allows the citizen to reach emergency services, despite the temporary speech impediment. The video footage provides vital information as to the nature and scale of the emergency and ensures that the most appropriate FRs are dispatched. The enhanced location capability of the NEXES App and the IP-enabled eCall service's location capabilities ensure that the emergency location is cross-referenced and determined in the swiftest possible time. Emergency system's networking ensures the linking of information between

the eCall of vehicle 2 and Person 1's emergency call and video footage, further enhancing the emergency services' situational awareness. FRs en-route have access to live intelligence on the emergency and the casualties' condition, ensuring all parties are well-informed, able to act rapidly and save lives.

Use Case 2: Chemical Plant Leakage

In the early hours of the morning, a tank at a small chemical plant develops a breach and hazardous materials begin leaking out. The security guard investigates the incident but fails to follow the appropriate security precautions and is quickly taken ill upon inhalation of noxious fumes. The guard attempts to call for assistance using his mobile device. Before being able to initiate the call, the guard loses consciousness.

Because the NEXES System has been adopted at the chemical plant, an automated emergency call via the NEXES telematics function is triggered following the detection of unusual and dangerous levels of chemicals in the air by installed sensors. Upon receiving this automated call, the emergency call operator has an immediate awareness of the emergency situation and rapidly dispatches the relevant personnel and resources, including specialist containment and decontamination crews and equipment.

Simultaneously, a second automated call is received at the emergency call-handling service. This call originates from the security guard's mobile device and has been triggered by the eHealth device detecting abnormal vital signs. The emergency call operator immediately attempts to make contact with the security guard through the smartphone. When they are unable to get a response from the caller, they use the device's location data sent in the automated call and dispatch an ambulance.

Despite being received at different emergency call handling centres, because both emergency calls have been managed through the next generation NEXES System they are now linked, thus ensuring that the different First Responders are aware of both the human casualty and the chemical hazard onsite.

Given the potential for broader environmental contamination and the adjacent location of a large residential area, a decision is quickly made to distribute location-specific push alerts via the NEXES System to the public, advising local citizens to stay indoors and seal their premises and providing an information number that residents may call for further information, ensuring the alerts do not result in an increased volume of emergency calls to emergency services.

The continued sharing of the environmental data via the NEXES telematics function ensures the firefighting FRs on site have real-time and accurate information on the extent and evolution of the hazard situation from the moment of dispatch, whereas the medical FRs are able to monitor the casualty's vital signs via their eHealth device whilst en-route to the scene. This information is fundamental in saving valuable time in the emergency response.

Use Case Highlights:
This *futuristic* scenario perfectly strengthens the potentially vital role of telematics in specific emergency scenarios. The use of telematics and automated calling means that in situations in which citizens using eHealth devices are incapacitated and unable to contact emergency services, they could be located and attended swiftly, with potentially life-saving consequences. Likewise, the capacity for emergency services to be contacted in the event of environmental hazards, whether in commercial or residential premises, could enhance the emergency response in terms of both response times and the quality and quantity of information available to them. Also important, the information received in real-time from hazardous sites can make a real difference in what respects to the safety of the emergency response teams on-site (adequate protection equipment). In this scenario, the NEXES push alerts function also enables emergency services to communicate the potential dangers to the public and advise citizens on the best course of action to ensure their safety.

Addressing real-world future application of IP-based technologies involving the use of telematics and smart devices (including eCall, health and environmental sensors) in emergency response, the NEXES use cases provide a valuable basis to validate the innovations envisaged by NEXES as a reference implementation of next generation emergency services.

3.2 *eCall*

A most notable and successful effort in bringing automated calls to next generation emergency services consists of the eSafety initiative of the European Commission, commonly known as eCall, duly present in NEXES.

eCall pioneered the integration of automated alerts and device data into the emergency services' information loop and it is a standardised in-vehicle caller device built upon the single European emergency number (112 and E112) that, in case of a serious accident, *automatically calls the nearest emergency centre. Even if no passenger is able to speak, e.g. due to injuries, a 'Minimum Set of Data' is sent, which includes the exact location of the crash site,*[6] *the time of incident, the accurate position of the crashed vehicle and the direction of travel. An eCall can also be triggered manually by pushing a button in the car, for example by a witness to a serious accident.*[7]

An eCall is recognised in the system and handled as a specialised form of emergency call, routed to a specialised eCall-proficient PSAP centre capable of processing the relevant set of vehicle and accident data (i.e., the MSD) has been standardised

[6] https://ec.europa.eu/digital-single-market/en/ecall-time-saved-lives-saved.

[7] http://ec.europa.eu/transport/themes/its/road/action_plan/ecall_en.

by the European Committee for Standardisation or CEN) and trained in handling emergency calls issued from vehicles. A pan-European pre-deployment of eCall was enabled through the EU co-funded projects HeERO and HeERO 2, leading to their ongoing successor I_HeERO that focuses on the upgrade of the Member States' PSAP software and hardware infrastructure, in order to be able to operate eCalls. In the HeERO projects, fourteen participating EU Member States and one associated country analysed the feasibility and robustness of the eCall service with pilot installations on both the PSAP and the vehicle.

In this prototype implementation, the In-Vehicle System (IVS) contains a GNSS module, offering the position of the accident that is embedded in a standardised compact emergency message (i.e., the MSD). The IVS contains also an In Band Modem to connect the IVS to the Public Land Mobile Network.

Upon the recognition of the benefits brought by the eCall telematics (2,500 lives per year expected to be saved throughout Europe with eCall[8]), on April 28th 2015, the European Parliament voted in favour of the eCall regulation, requiring all new cars to be equipped with eCall technology from April 2018 [13]. Still, today no full integration of eCall modules has been accomplished in PSAP centre environments.

The original eCall concept was developed considering emergency calls originated from circuit-switched (analogue) mobile telecommunications infrastructures, specifically 2G and 3G networks using an in-band modem. As technologies evolve and NGES adopt a 4G communications framework - fully digital IP-based [12] - the current eCall concept needs to be refined and adapted, so as to benefit from the resource-efficient packet-based systems

Starting from reference use case 1, the NEXES RIA has defined the major architectural features, the high-level application components, the key functional requirements and the required layout and process flow to perform the implementation of the NG eCall service.

Next Generation IP eCall: the eCall Evolution

As pointed by the European Emergency Number Association (EENA) in the technical report on NG eCall,[9] "*circuit switched 112 will eventually be replaced by IMS emergency call in UMTS-PS and LTE networks and it is further expected that EU operators will want to phase out support of circuit-switched (CS) GSM and UMTS over the next decade or two. […] The issue of how in-band modem eCall deployments in cars will continue to be supported, whilst network technology migrates to IMS, requires consideration. […] As IMS emergency call capability will be extensively deployed well before the 2030s, in-band modem eCall and IMS eCall will have to co-exist.*"

Recent advances in personal smart devices technologies such as smartphones and tablets are making it possible to detect car accidents in a more portable and cost-effective manner than conventional in-vehicle solutions. In addition, car adapter products are already out there for connecting vehicle diagnostic data to the user's

[8]http://www.etsi.org/news-events/news/960-2015-05-european-parliament-makes-ecall-mandatory-from-2018.

[9]http://www.eena.org/download.asp?item_id=162.

smartphone through the On Board Diagnostic (OBDII) vehicle gateway and may also include an emergency service subscription. The pervasiveness of smartphones also means that the infrastructure required to establish a wireless mobile sensor network is already in place and available after installing appropriate application software. Smartphone manufacturers also have begun including a plethora of sensors that enable devices to detect the context in which they are being used. For example, an Android-based device that possesses a compass, accelerometer and GNSS receiver allows application developers to determine the geographic position, heading and movement of the user. The processing power, popularity and relatively low cost [5] make smartphones an appealing platform to construct a wireless mobile sensor network that detects car accidents.

Extending the original eCall specifications [14], NEXES aims to implement a next generation eCall prototype that embeds the eCall features into the 4G communication framework and introduces novel functionality in the communication between the IVS and the PSAP (going beyond MSD exchange and PSAP call back features) as envisioned and specified in the Next Generation Pan-European eCall draft standard drafted by IETF [12]. A NEXES prototype will be based on fully programmable open interfaces and an Application Programming Interface (API) being suitable for deployment in both traditional IVS on-board systems and smart devices acting as IVS.

Work on NG eCall (also referred to as packet-switched eCall) is ongoing with the firm objective to re-use the emergency services' mechanisms specified for SIP to accomplish emergency call support in vehicles [15]. As part of this effort, the European Telecommunications Standards Institute (ETSI) has published a Technical Report titled *Mobile Standards Group (MSG); eCall for VoIP* [16] that presents findings and recommendations regarding the support for eCall in an all-IP environment. eCall is hence expected to gradually evolve to NG eCall as circuit-switched environments are replaced by all-IP ones, bringing forth vehicle data and other eCall-specific data as additional data, associated with the emergency call.

Based on the Next Generation Pan-European eCall specification, NEXES identified relevant high-level requirements to be satisfied by the eCall system, including those that are specific to a next generation eCall. They were originally presented in [17] and are presented in Table 2.

Architecture of the NG eCall

The high-level architecture of the NG eCall service is based on the legacy eCall system [14, 18] and the need to consider modern UE, such as smartphones. Applicable components thus comprise the client system and the eCall module in the PSAP system.

As depicted in Fig. 3, and following the work developed in the European HeERO research project, the **NG eCall client system** combines two main modules: the **Vehicle Gateway**, responsible for gathering the relevant vehicle data (originated by the vehicle's internal and external sensors), and the **Client Device**, responsible to trigger, manage and effect the eCall transaction, including preparing and sending the MSD.

Table 2 Main functional requirements of current and next generation IP eCall

eCall	NG eCall added features
The call is recognised as an eCall (inherently an emergency call)	The call is able to carry more data (e.g., an enhanced Minimum Set of Data or an MSD plus additional sets of data)
The call setup indicates if the call is manually or automatically triggered	The call is able to handle video
The call is able to provide a voice channel between the vehicle and the PSAP	The call is able to handle text
Carries the MSD intrinsically with the call (the MSD needs to be available to the PSAP operator as the voice call)	The PSAP is able to access vehicle components (e.g., on-board cameras used for parking for a visual assessment of the crash site situation)
The PSAP is able to acknowledge receipt of the MSD	The PSAP is able to request the vehicle to take actions (e.g., sound the horn, disable the ignition, lock/unlock doors)
The PSAP is able to request that the vehicle generates and transmits a new MSD	The call is able to simultaneously handle voice and data exchange
The PSAP is able to call back the occupants of vehicle after the initial eCall is terminated	
Supports a test call (which can be routed to a PSAP but is not treated as an emergency call and not handled by a PSAP operator)	

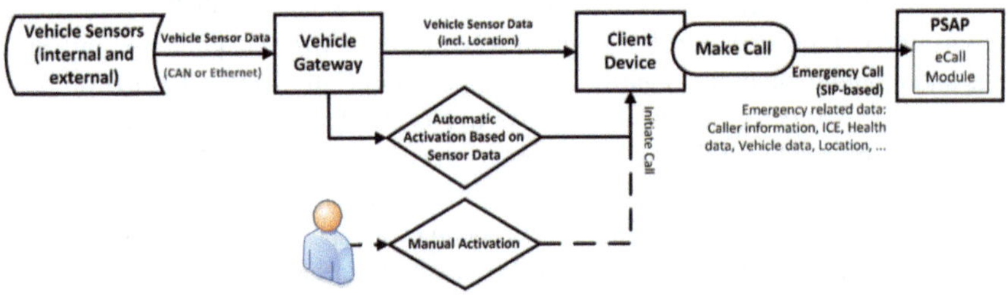

Fig. 3 NG eCall high-level architecture

The **Vehicle Gateway** is responsible for collecting the vehicle's internal (e.g. speedometer) and external (e.g. accelerometer device) sensor data and is able to indicate a *crash* condition based on these data values. It is usually a small vehicle industrial personal computer that interfaces with the vehicle sensors.

The **Client Device** may be a built-in IVS or a smartphone supporting IP connectivity and it is the component responsible for placing the NG eCall. Operation is assumed to be limited within a SIP-based environment, as per specifications in 3GPP's Technical Specifications TR 26.967 V8.0.1 [11]. The Client Device receives inputs (e.g., vehicle sensor data) from the Vehicle Gateway and hosts the NG eCall Application, i.e., a software that determines whether an eCall should be automatically initiated. The location is provided either from the Vehicle Gateway or directly from the Client Device (i.e., an IVS equipped with GNSS or a GNSS-equipped smartphone or a location acquired from the smartphone's mobile network).

Residing in the user's smartphone or in the IVS equipment, the NG eCall Application is thus responsible to process the sensor (collected vehicle data, including a crash indication) and user (selection of voice, text or video) inputs, presenting user interface routines that should at least include a simple visual/audio interface where (i) the user may initiate an eCall (selecting NGES functions such as audio, video and real-time text) and (ii) the system may interact with the user to inform that an automated eCall was initiated. The NG eCall Application then uses a data communication channel, enabling TC and the eCall data exchange, to provide emergency call-related data to the PSAP centre, during an emergency call in an all IP environment. As per the Pan-European Mobile Emergency Application (PEMEA) recommendation, the data channel is offered by a SIP/VoIP provider (on the client side) and a PSAP Service Provider (on the PSAP side) [19]. For purposes of interoperability with legacy PSAP systems, a voice-only channel is also included to support communications [12, 19].

Considering the NG eCall activation, NEXES considers both manual and automated activation modes, following the IETF recommendation [20]: automatically triggered eCalls indicate a car crash or a serious accident and carry greater presumption of risk of injury since no human intervention has taken place; and user-initiated (manually-triggered) eCalls might be reports of accidents, witnessed crashes or serious hazards but they are also more likely to be accidental calls [12]. While for the latter, human confirmation in call back is possible, in the former, it may not be the case, thus PSAPs might apply different procedures in handling automated and user-initiated eCalls. These two modes of activation are also depicted in Fig. 3.

Figure 3 furthermore includes the **NG eCall module in the PSAP system**, an example of the many adaptations and upgrades existing PSAP systems are undergoing to capture the opportunity brought by IoT, automation and the new IP-enabled

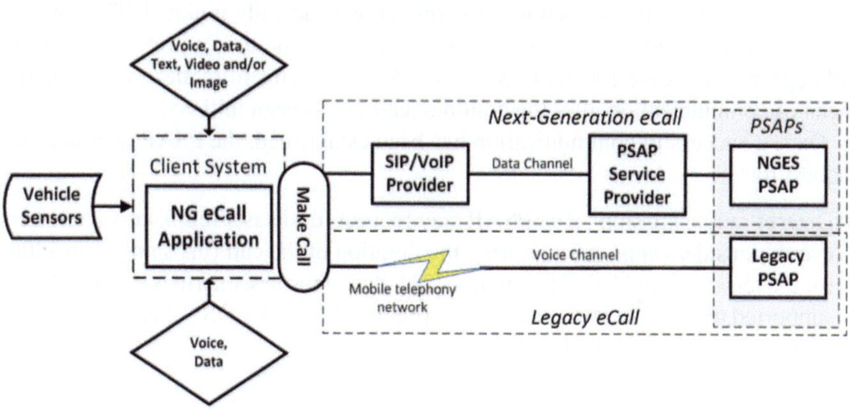

Fig. 4 NG eCall application overview (*source* [17])

technologies and integrate more and new types of data in the information exchange between citizens and PSAPs in emergency situations (Fig. 4).

NG eCall State Flow

Adapting the legacy eCall specifications [14] for NG environments, the NG eCall state flow between the client device (herein, the IVS equipment) and the eCall Application in the PSAP handling the eCall comprises eight steps:

- Step 1 – Procedures following power-up and initialisation of the IVS;
- Step 2 – Activation of the system;
- Step 3 – NG eCall set-up, including identification of call type, make call, network selection and registration, authentication (home location registry), cell location by network, establish connection to PSAP centre (voice, video or RTT);
- Step 4 – NG eCall dataset transfer;
- Step 5 – Establish voice or RTT or video link;
- Step 6 – Clarification of the overall emergency situation;
- Step 7 – Initiate accident resolution and inform vehicle occupants verbally that help is coming;
- Step 8 – NG eCall termination.

Figure 5 provides the NG eCall state flow adopted in NEXES, based on the Use Case No 1, where the states are used for sequential design to create state transition sequences, the processes are represented with boxes and the control blocks are denoted with rhombus. It is worth noting that steps 1 and 2 above-described are now considered obsolete, as well as step 7, included in the *Successful Communication* control block. The term *Dataset* represents NG eCall extended set of data, comprising both vehicle data and user data, as defined in [20].

NEXES assumed that once the NG eCall is triggered, the NG eCall Dataset is transferred and, sequentially, communication with the user is attempted. Although this process is not mandated by SIP, it is deemed more efficient to have the PSAP operator receiving the user data before the communication is actually initiated. The communication means used (voice, video or RTT) abide to the user's preference/profile, although it is possible for the user or the PSAP operator to request an alternative mean of communication after the communication has been initiated.

Once a successful communication has been established, the PSAP operator may request for:

- Dataset retransmission - The PSAP may request for dataset retransmission whenever it considers appropriate during the duration of the call (the same is specified for the legacy eCall [13]). Note that in [21], the opposite scenario is proposed to be supported too: "If the IVS is aware that previously sent VEDS[10] data has changed, it may send an unsolicited VEDS during the call".

[10]The Vehicular Emergency Data Set (VEDS) is a XML-based data standard that determines useful and critical elements needed for an efficient emergency response to vehicular emergency incidents. The Protocol identifies crash and medical data elements. The full VEDS data definition can be found in [18].

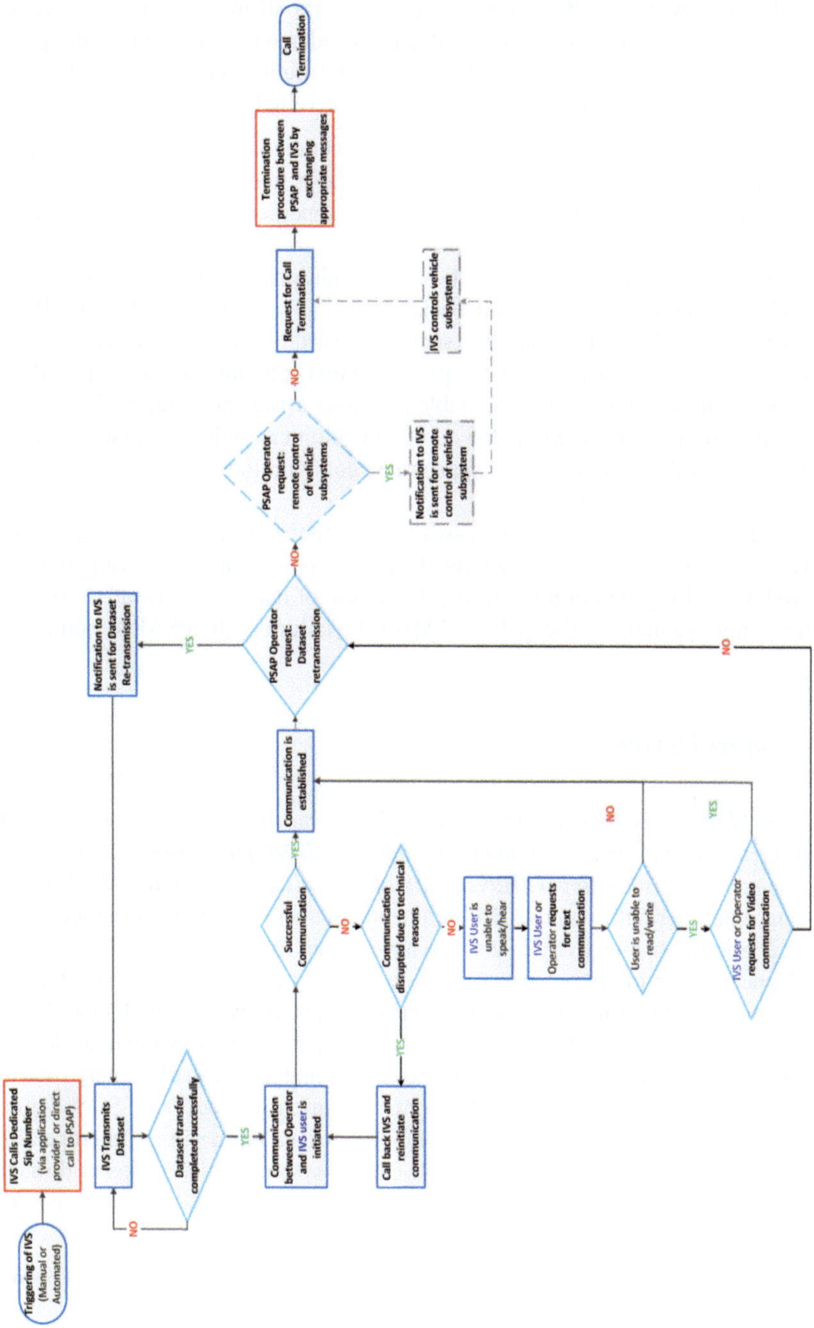

Fig. 5 NG eCall state flow (*source* [17])

- Remote control of vehicle subsystem (optional – indicated with dashed lines in Fig. 4) - Following the draft standard [21], the PSAP may optionally[11] request access to control a vehicle subsystem that is considered relevant for the emergency response such as turning on the flashers or the lights, remotely unlock doors, honking the horn or enabling an exterior/interior vehicle camera.

In order to enable such options, the possible actions and data types relevant to vehicle indicators and subsystems (such as available lamps and cameras) that are supported by the vehicle should be included in the emergency dataset transmitted by the IVS.

When communication fails, the following conditions are checked sequentially: the PSAP operator attempts to call back (voice call); if unsuccessful, the PSAP operator assumes the user is unable to speak or hear the call and requests that a text communication is established. If the request for a text communication is rejected, the PSAP operator assumes the user is unable to read or write and requests that a video communication is established. Video communication can only take place upon the prior consent of the user or/and the appropriate configuration of the vehicle, in case of a built-in camera system.

Procedures are also required for cases where other failures occur, including the incorrect transmission of the dataset, the dataset not being sent, the dataset not being received, the false generation of eCalls, the failure of network registration, the NG eCall routing to a non-eCall-capable PSAP or the failure of the PSAP system.

3.3 Smart Devices

"The world of IoT is currently characterised by competing technologies and platforms, further complicated by numerous standards development organisations, and this fragmentation is causing a delay in the widespread adoption of IoT.".[12] Notwithstanding, interoperability among different standards and communication technologies [22] and the connectivity landscape have begun to change slowly, aided with the advancement in cloud services [23] by technology leaders such as Amazon and Google and the beginning of the *smart home* concept deployment, with Google and Samsung introducing smart hubs to be interoperable with a variety of smart devices increasing the potential number of connected nodes that users can ultimately control.

The Open Geospatial Consortium Sensor Observation Service (OGC-SOS) defines standard web service protocols for sharing sensor data online in an interoperable manner. Still, OGC-SOS has a relatively weak ecosystem, which makes it difficult to build and consume, and it only supports predefined queries. On the contrary,

[11] Remote control concept described in here assumes the regulatory framework in existence permits the requested actions to be performed from the PSAP side. The concept is presented as optional (dashed lines in the diagram) since this is not the current situation where vehicle internal communication network as well as vehicle subsystems' use is proprietary by car manufacturers.

[12] http://www.tetratoday.com/news/smart-cities-smart-emergency-services.

the Open Data Protocol (OData) built by the Organisation for the Advancement of Structured Information Standards (OASIS) presents a strong ecosystem and flexible query functions. But the soft-typing approach of OData requires it to have a commonly agreed data model to be interoperable. Hence, Chih-Yuan Huang and Steve Liang proposed a sensor data mediator solution and defined a SOS entity data model for OData (SOS-OData) to bridge these two standards and promote the idea of an *open data for sensor web specification* [24].

In NEXES, state-of-the-art signaling mechanisms (such as SIP) and protocols for sensor data conveyance to PSAPs will be studied, leading to two prototype implementations for smart building and personal health monitoring in the NGES context, sharing where appropriate similar principles with the NEXES eCall implementation.

It is ambitioned that NGES systems incorporate smart devices, such as sensors, smartphones and tablets, and explore their applicability to initiate emergency calls, either automatically or upon user initiative, and transfer relevant data to emergency services to improve emergency response efforts.

Sensor data envisaged as very relevant in emergencies and ambitioned to be included in the NG emergency information loop includes data from environmental sensors (e.g. temperature sensors, accelerometers) installed in specific building locations. Smart buildings are bound to contribute to the creation of monitored grids across smart cities and their role in the emergency context is significant since monitoring will effectively sense critical environmental conditions, namely fire spreading or earthquakes. Other data deemed important in emergencies are the data from health sensors, i.e. from wearable devices on patients with known health problems or citizens experiencing special needs, as well as the data from integrated sensors embedded in smartphones and tablets, such as accelerometers or gyroscopes.

In the context of health sensors and devices, smartphones or tablets would act as intermediate gateways to collect the sensors data through a smartphone application and send it via wireless links, utilising available technologies, such as 4G or WiFi, to a web server application.

In NEXES, two sensor-based NGES applications illustrated in the Use Case No.2 will be investigated and explored: a smart building application and a smart health application.

Depicted in Fig. 6, the **smart building application** delivers a monitoring capability over the building floors based on the data provided by a combination of sensors comprising smoke detectors, accelerometers, thermometers and gas sensors.

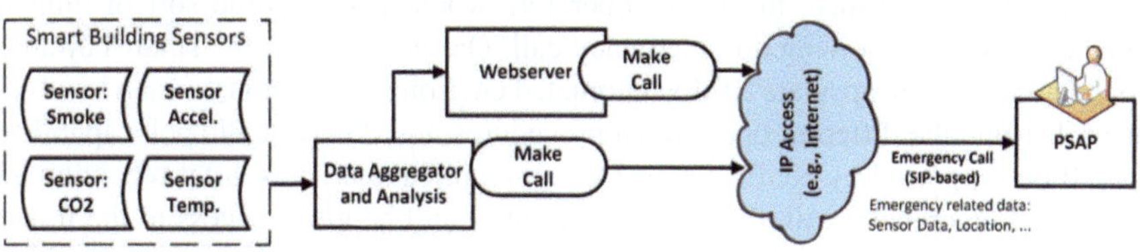

Fig. 6 Smart building application (*source* [17])

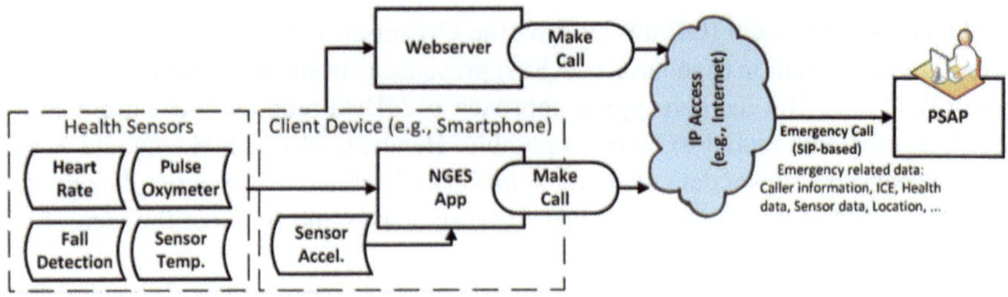

Fig. 7 Smart health application (*source* [17])

Sensor data is aggregated by a *Data Aggregator and Analysis* component and it may be applied to indicate an emergency situation. If deemed appropriate, this component may initiate an automated emergency call. In alternative, the automated emergency call may be established by the *Webserver* component, deployed to collect the relevant data from the *Data Aggregator and Analysis* component. This alternative may be particularly useful in case the *Data Aggregator and Analysis* component does not have the capability – or is not intended – to initiate emergency calls or in situations where a third party is responsible for detecting anomalies and initiating (manually or automatically) emergency calls. In this setting, location information details are predefined in each sensor (assumed fixed), allowing the PSAP operator to accurately determine exactly where (i.e., address, floor and room) the incident is located.

The **smart health application**, illustrated in Fig. 7, acts as a personal health monitoring application, recording the user's health parameters by means of sensors connected to the user's smartphone. In addition, smartphone sensors, such as the accelerometer for fall detection, are also considered when collecting the user's health data. A NG emergency App, residing in the user's smartphone, analyses the collected sensor data and may initiate an automated emergency call if abnormal (life threatening) values are detected.

Similarly to the smart building application, a *Webserver* component can be deployed to collect relevant data from the user's smartphone (or multiple users' smartphones) and be used instead to initiate an automated emergency call, which may be particularly useful in case a third party is involved in the detection of anomalies and the (manual or automated) initialisation of the emergency call. Location information is also collected using the smartphone's capabilities.

It should be noted that, because the personal health monitoring application does involve human presence, the PSAP operators would expect some sort of human interaction when receiving an emergency call. On the contrary, the smart building application may encapsulate a fully automated environment without human involvement. Despite the different operational procedures, the PSAP's call back capability should be supported by NGES Smart Devices applications: thus, after receiving an automated emergency call, the PSAP operator would be given access to the transferred emergency-related data (e.g., query the web server component and collect relevant details), in order to improve the situational awareness and act accordingly.

NGES Smart Devices State flow

In Fig. 8, a NGES smart devices' state flow presents the establishment of an automated emergency call, where states are used for sequential design to create state transition sequences, processes are denoted with boxes and control blocks are symbolised by rhombus.

A *Gateway* component is introduced to represent a generic device-agnostic responsible of initiating a NG emergency call. The *Gateway* component can be the *Data Aggregator and Analysis*, the *Smartphone* or the *Web Server* components described earlier. The applications' operation is assumed to be limited to a SIP-based environment and the applicable signalling mechanisms (e.g. message encapsulation inside SIP, call back mechanisms) are outside the scope of the state flow.

The state flow displays the presentation of several conditions to ensure a successful communication is established: the PSAP operator may request for Dataset retransmission, the PSAP operator may request for activation of auxiliary devices and the PSAP operator may access and modify data transmission configuration (e.g. increase data sampling). Moreover, specific procedures should also be defined for failure situations, namely the incorrect transmission of the Dataset, the false generation of automated smart devices emergency calls, the failure to register the network, among others.

3.4 *Proposed Data Structures for Automated NG Emergency Calls*

Whether an automated emergency call is established by a NG eCall service, a smart building application or a smartphone's sensors or personal health monitoring application, the additional data it generates in relation to the emergency and transmits to the PSAP operator can have a significant impact in the emergency services' situational awareness and dramatically improve the efficiency, effectiveness and outcome of the emergency response effort.

In an attempt to define a common and harmonised format for exchanging NG emergency call additional data, extensive work has been conducted by several institutions and standardisation bodies, including 3GPP, CEN, EENA, IETF, ETSI and NENA [25]. NEXES has studied the work conducted and highlights the main information sources and the relevant data structure elements adopted in its reference implementation of a NGES (Table 3).

In addition, NEXES also determined the additional data categories required to support automated NG emergency calls, identified in Table 4, along with the references to the relevant standards.

With regards to the data transmission in the NG Emergency Calls, NEXES acknowledges that the following applies:

- The mechanisms described in [12] allow the additional data to be conveyed by value (within the body of a SIP message or a location object) or by reference

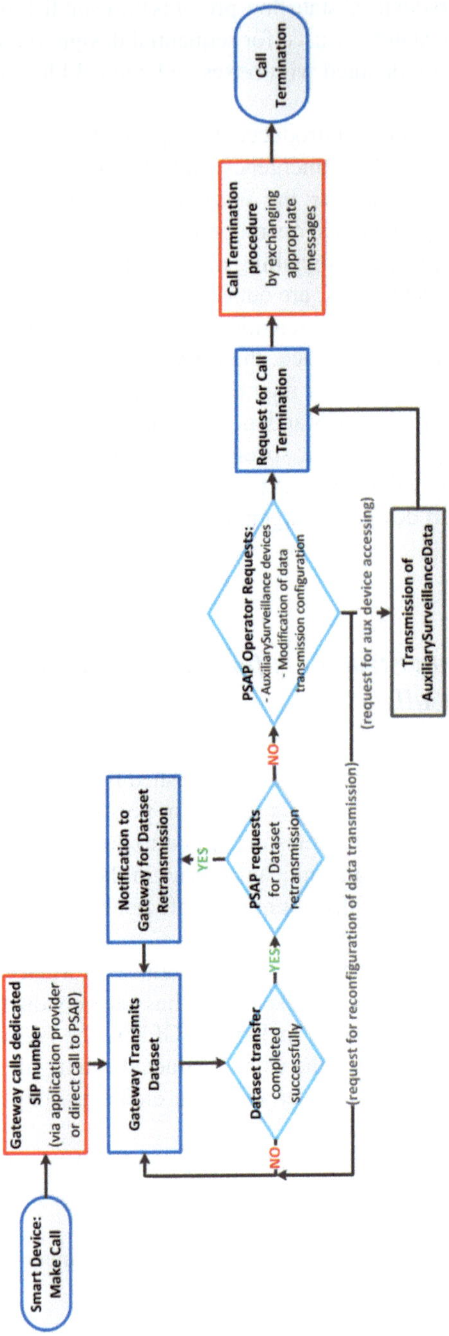

Fig. 8 Smart sensor emergency call process flow (*source* [17])

Table 3 Adopted data structure elements and applicable sources

Reference	Data structure elements adopted
Additional Data Related to an Emergency Call draft-ietf-ecrit-additional-data-37 [20]	Elements related to user and service (core message elements related to "a Call" as defined in [12])
Next-Generation Vehicle-Initiated Emergency Calls draft-ietf-ecrit-car-crash-09 [26]	Vehicular Emergency Data Set (VEDS) Recommendation (March 2004) [27]: The Vehicular Emergency Data Set (VEDS) is a XML-based data standard that determines useful and critical elements needed to prove an efficient emergency response to vehicular emergency incidents. The Protocol identifies crash and medical data elements. The full VEDS data definition can be found in [27]. NEXES adopted the:
	– VEDS data elements defined in [12] related to a car crash (to enrich the Vehicle Data category);
	– Medical profile data.
IETF's RFC5491 [27]	Location
IETF's RFC7459 [28]	Location confidence
CEN - EN 15722 [29]	Legacy eCall MSD data specifications

(as an external resource). This feature follows the tradition of prior emergency services standardisation work, where data was conveyed by value within the call signaling (i.e., in the body of the SIP message) or by reference.

- *The IETF's Additional Data related to an Emergency Call* document [20] establishes a general mechanism for attaching blocks of data to a SIP emergency call. However the size and frequency of additional data transmission during a session needs to be evaluated to ensure its appropriateness to use the signaling channel over the inclusion by reference.
- The vehicles designed for multiple regions might need to support NG eCall and other Advanced Automatic Crash Notification systems, such as described in [26].

3.5 The Integration of Automated NG Emergency Calls in PSAP Systems

The PSAP operators are the voice and face of emergency services that citizens first interact with when reaching to emergency services in emergencies. Their importance in the emergency services' value chain is therefore obvious and the best outcome of their efforts is directly related with the performance of the system that support their

Table 4 Data categories to support automated NG emergency calls (*source* [17])

Data category	Description	Reference
Data provider	This block supplies name and contact information for the entity that created the data	[20], p. 9
Service information	This block supplies information about the service	[20], p. 17
Device information	This block supplies information about the device placing the call	[20], p. 22
Owner/subscriber	This block supplies information about the owner of the device or about the subscriber	[20], p. 27
Short message	This block provides a way to supply free form human readable text to PSAPs or emergency responders	[20], p. 31
General data	This block supplies general information, such as message id and current location	[28–30]
Vehicle data	This block supplies information regarding the vehicle from which the call was originated in case of automotive applications	[25, 26, 29]
	Note 1: Contains parts of crash related data as defined in VEDS [27]	
	Note 2: The investigation for vehicle related 43268data suitable for emergency response was guided by [12].	
Health data	This block supplies medical profile data regarding the owner of the device or the subscriber	[27]
	Note: Contains parts of medical profile data as defined in VEDS [27]	
Sensor data	This block supplies sensor data resulting from the detection of an emergency event by an eCall or a smart device. Events can consist of e.g., heart rate high variability, fall detection, crash detection, smoke detection and CO^2 level overstepping detection	NEXES [5]
Additional data	This block can be used to transmit additional data (such as specific sensor data in case of automotive applications) or undefined data	[27], p. 16

service. PSAP systems assist emergency response from integrated call centre activities to incident dispatch, providing services that include the search of addresses and locations, the identification of the accurate location of incidents, constant information updating, the dispatching of resources and interventions and the interoperability and coordination of agencies during emergency interventions.

Nowadays, PSAP centres are faced with the pressing need to update their systems in order to provide an efficient and effective response to the many solicitations of citizens using new technologies to reach emergency services, as well as to the challenges posed by the integration of telematics and smart devices with emergency services through the avenue of automated NG emergency calls, following the European Parliament's stance on mandating that new vehicles be equipped with the eCall technology from 2018.[13]

Indeed, automated NG emergency calls are directly related with the uptake of total conversation capabilities, smart device data and location-aware devices and the advent of telematics and smart devices in the emergency information loop. The integration of automated NG emergency calls in PSAP systems determine that the NG PSAP systems have reinforced capabilities to handle additional communication channels, including video and real time text, as well as a larger volume of data (e.g., enhanced MSD and additional sets of data) that will include details on the data provider, service information, device information, vehicle information, owner/subscriber information, timestamps, location-related data, sensor data and health data.

When an automated or data-enabled IP-based (NG) emergency call is initiated, a rich set of data from multiple data sources can be conveyed to the PSAP centre. This data includes (but it is not limited to) information about the multimedia capabilities of the device, the request for emergency services, location information and metadata about the nature of the emergency and the data sources. IETF's document on Additional Data Related to an Emergency Call (a proposed standard) extends the basic set of data communicated with an emergency call [31] based on SIP, as described in IETF's *Framework for Emergency Calling Using Internet Multimedia* [32] and *Best Current Practice for Communications Services in Support of Emergency Calling* [33] documents, in order to carry different types of additional data that is useful for the PSAP operator to handle the next generation emergency call. Attentive to these standardisation efforts, also NEXES proposes to integrate automated NG emergency calls in the NEXES PSAP system using SIP-based interfaces that benefit from the requirements of NG services. In addition, the NEXES PSAP System will provide new functionality, enabling interoperability among different PSAP centres, through the support to emergency calls forwarding to other PSAPs, deemed more appropriate in function of required and available resources, technological equipment, language services and workload status.

Aside from the integration of new functionality and the specific concern with the new rules of prioritisation for innovative TC emergency calls over NG IP

[13] http://www.europarl.europa.eu/news/en/news-room/20150424IPR45714/automatic-emergency-call-devices-in-all-new-car-models-from-spring-2018.

networks, PSAP systems also have their traditional features quite strained. One of these features is the call back function. Prompted by the premature termination of the emergency call before the PSAP operator collects all the required emergency-related information, the call back function empowers PSAPs to reconnect to the emergency caller using the contact data previously provided with the initial emergency call. In situations involving automated NG emergency calls, initiated by smart sensors, smartphones or the eCall device, the call back functionality brings an added-value, not only confirming the early detection of an emergency (thus, contributing to a better outcome) but also empowering the opportunity for the PSAP system to remotely access specific sensors (e.g., cameras on-site, specific vehicle equipment, sensor platforms), upon prior consent given by the user, to improve the PSAP operators' situational awareness with respect to the emergency and the adequate emergency response: capture images of the incident site, turn-off vehicle's equipment, change the sensors' data exchange frequency to retrieve additional data and perform better evaluation of the emergency.

With the integration of richer and additional sensor data into the emergency information database of PSAP systems, these systems need to reinforce the robustness and reliability of relevant system support functions, including session recording, audio and multimedia recording, workstation registration, indexing and tagging, reporting, data fusion and processing, so as to prevent information conflict and overload. In NEXES, the PSAP system displays all the emergency-related information to the PSAP operator in an intuitive and easy-to-use form, overlaying maps, displaying easy colour schemes and presenting rich information windows for improved decision-making and minimal operational impact.

Moreover, a new attention is given to the different facets of information security, so as to guarantee and protect the confidentiality, integrity and availability of assets in PSAP centres, ensuring their services' quality and sustainability. As NGES are expected to heavily rely on Internet-based technologies and data exchange, technologies that are exposed to a variety of threats to the availability, confidentiality and integrity of the NGES' data and resources. New threats, exploitation modes and vulnerabilities emerge and may prevent the optimal performance of the new capabilities and the general credibility of the PSAP service provided. Attacks include Distributed Denial of Service, man-in-the-middle attack and identify theft. PSAP systems need to reinforce their protection and security measures, such as encryption, authentication and firewall mechanisms, to handle situations where, for example, malicious Apps and smart devices are triggered at a certain date and time to initiate fake emergency calls that flood PSAP centres and compromise service availability. Requiring that Apps and smart devices accessing NGES are certified and supplied by trusted sources will certainly decrease the likelihood of misbehaviour and the opportunity to foul play. Indeed, providing harmonised response from the whole emergency services' value chain to these new and emergent security challenges are the best form to ensure and safeguard the emergency services' capability to fulfil their mission.

4 Conclusion

As society begins to evolve towards next generation emergency services, the use of telematics and smart devices in emergencies emerges amongst the potential applications with the most promising impact and benefits.

The capability to empower the early detection of emergencies, including in situations where human intervention might not be possible or is impaired, as well as an enhanced emergency situational awareness, benefitting from the real-time exchange of relevant information, are all the more reasons to adequately explore, implement and validate the embrace of automated NG emergency calls worldwide. Simply put, the adoption of the IoT paradigm by NG emergency services has the potential to revolutionise the quality of emergency response and save lives in the process.

Indeed, this acknowledgement has been undoubtedly expressed by the European Union, when the European Parliament voted in favour of the eCall regulation on April 28th 2015 and required that all new cars be equipped with eCall technology from April 2018. It is expected that the eCall telematics saves 2,500 lives per year throughout Europe.

Inspired with the new challenge of adopting the IoT paradigm to the emergencies domain and create a reference implementation of a next generation emergency service, the NEXES RIA explores the benefits, shortcomings and challenges involving the integration of telematics and smart devices in emergency services, leveraging on three key enablers: total conversation, device data and location-aware devices. NG eCall, eHealth applications and smart environmental sensors constitute the platforms to concept-prove and showcase the added-value generated by automated NG emergency calls for a cost-effective, efficient and high-performing emergency response. Relevant information sources and data structure elements have been thorough analysed and the latter adapted to handle additional sensor data and define a framework for automated NG emergency calls that convey true innovation on new additional data structures capable of addressing the specificities of automated and data-enabled emergency calls.

Nonetheless, the acceptation of automated calls by emergency services still faces obstacles, especially considering that IoT is at its infancy and smart devices' reliability and security aspects still need to be carefully addressed and improved.

As an exploratory research endeavour, NEXES embarks in this journey. And, as it begins to develop and to create an end-to-end NGES solution, technical, technological, operational, standardisation, ethical and regulatory challenges will continue to emerge, a constant reminder of the overwhelming importance of engaging all relevant stakeholders in the emergency services value chain, if automated NG emergency calls are to become a reality across Europe.

Acknowledgements This work has been prepared as part of the NEXES Research and Innovation Action, which has received funding from the EU's Horizon 2020 research and innovation programme under Grant Agreement No. 653337.

References

1. Atzori, L., Iera, A., Morabito, G.: The Internet of Things: A Survey, Computer Networks Journal, 31 May 2010. https://www.researchgate.net/profile/Luigi_Atzori2/publication/222571757_The_Internet_of_Things_A_survey/links/546b36df0cf2f5eb180914e5.pdf
2. Fortino, G., Guerrieri, A., Lacopo, M., Lucia, M., Russo, W.: An agent-based middleware for cooperating smart objects. In: PAAMS (Workshops), pp. 387-398 (2013)
3. European Emergency Number Association, Next Generation 112 Long Term Definition, EENA NG 112 Document, 11 Apr 2012. http://www.eena.org/uploads/gallery/files/pdf/2013-03-15-eena_ng_longtermdefinitionupdated.pdf
4. European Emergency Number Association, Public Safety Answering Points (PSAPs) in Europe, EENA, 13 Nov 2014. http://www.eena.org/uploads/gallery/files/pdf/PSAPs_in_Europe_2014_Abstract.pdf
5. NEXt generation Emergency Services (NEXES). http://nexes.eu
6. Coordination Group on Access to Location Information for Emergency Services, Final report on implementation issues related to access to location information by emergency services (E112) in the European Union (2002)
7. European Commission, The European Emergency Number 112, Flash Eurobarometer 368. http://ec.europa.eu/public_opinion/flash/fl_368_en.pdf (2013)
8. Palen, L., et. al.: Crisis informatics: studying crisis in a networked world. In: Third International Conference on e-Social Science. Michigan, 7–9 Oct 2007. www.cs.colorado.edu/~palen/palen_papers/palen-crisisinformatics.pdf
9. Rosen, B., et al.: Additional data related to an emergency call, internet-draft, internet engineering task force, 23 Apr 2014. http://tools.ietf.org/html/draft-ietf-ecrit-additional-data-22
10. Manso, M., Guerra, B., Amditis, A., Bolovinou, A., Sdongos, E., Carjan, C., Donaldson, D.: The application of telematics and smart devices in emergencies: use cases in next generation emergency services. In: IEEE International Conference on Cloud Engineering (IC2E), Berlin, Germany, 4–6 Apr 2016
11. 3rd Generation Partnership Project, Technical Specification Group Services and System Aspects, eCall Data Transfer; In-band modem solution (Release 8), TR 26.967 V8.0.1. http://www.arib.or.jp/IMT-2000/V720Mar09/3_T12/ARIB-TR-T12/Rel8/26/A26967-801.pdf (2007)
12. Gellens, R., Tschofenig, H.: Next-Generation Pan-European eCall, draft-ietf-ecrit-ecall-07.txt, The Internet Engineering Task Force, 19 Feb 2016. https://tools.ietf.org/html/draft-ietf-ecrit-ecall-07
13. Regulation (EU) 2015/758 Of The European Parliament And Of The Council of 29 April 2015 concerning type-approval requirements for the deployment of the eCall in-vehicle system based on the 112 service and amending Directive 2007/46/EC. http://eur-lex.europa.eu/legal-content/EN/TXT/?uri=uriserv:OJ.L_.2015.123.01.0077.01.ENG
14. BS EN 16062:2015, Intelligent transport systems. eSafety. eCall high level application requirements (HLAP) using GSM/UMTS circuit switched networks, British Standard (private copy)
15. Rosen, B., et al.: Internet Protocol-based In-Vehicle Emergency Call, draft-rosen-ecrit-ecall-06.txt, The Internet Engineering Task Force, 15 July 2012. https://tools.ietf.org/html/draft-rosen-ecrit-ecall-06
16. European Telecommunications Standards Institute, Mobile Standards Group (MSG); eCall for VoIP, ETSI TR 103 140 V1.1.1. http://www.etsi.org/deliver/etsi_tr/103100_103199/103140/01.01.01_60/tr_103140v010101p.pdf (2014)
17. Sdongos, E., Bolovinou, A., Tsogas, M., Amditis, A., Guerra, B., Manso, M.: Next generation automated emergency calls. In: IEEE CCNC (Globe-IoT, 2017). Las Vegas, Nevada, USA, Jan 8–11 2017
18. Thomson, M. et al.: Relative Location Representation, RFC7035, The Internet Engineering Task Force (2013)

19. Winterbottom, J., et al.: Pan-European Mobile Emergency Application (PEMEA) Requirements and Functional Architecture, EENA NG112 Technical Committee Document, European Emergency Number Association. http://www.eena.org/download.asp?item_id=158 (2015)
20. Gellens, R., et al.: Additional Data Related to an Emergency Call, The Internet Engineering Task Force. https://tools.ietf.org/html/draft-ietf-ecrit-additional-data-37 (2015)
21. Winterbottom, J., et al.: Specifying Civic Address Extensions in the Presence Information Data Format Location Object (PIDF-LO), RFC 6848, The Internet Engineering Task Force (2013)
22. Aloi, G., Caliciuri, G., Fortino, G., Gravina, R., Pace, P., Russo, W., Savaglio, C.: Enabling IoT interoperability through opportunistic smartphone-based mobile gateways, J. Netw. Comput. Appl. (to appear)
23. Fortino, G., Guerrieri, A., Russo, W., Savaglio, C.: Integration of agent-based and Cloud Computing for the smart objects-oriented IoT, CSCWD, pp. 493–498 (2014)
24. Huang, C., Liang, S.: A sensor data mediator bridging the OGC Sensor Observation Service (SOS) and the OASIS Open Data Protocol (OData), Ann. GIS 20(4):279–293 (2014). doi:10.1080/19475683.2014.942795
25. Irwin, D.: Automatic Collision Notification and Vehicle Telematics Technical Information Document (TID), 07-504, NENA Technical Information Document. National Emergency Number Association, USA, 1 June 2007. https://c.ymcdn.com/sites/www.nena.org/resource/collection/91E03A3A-B334-4EB6-9205-E25BBF6AF8E7/NENA_07-504-v1_Collision_Notification_and_Telematics_Information.pdf
26. Gellens, R., et.al.: Next-Generation Vehicle-Initiated Emergency Calls, draft-ietf-ecrit-car-crash-09.txt, The Internet Engineering Task Force, 1 Aug 2016. https://tools.ietf.org/html/draft-ietf-ecrit-car-crash-09
27. ComCARE Alliance, Vehicular Emergency Data Set (VEDS) Recommendation, ACN Data Set Working Group, Version 2.0. http://xml.coverpages.org/ComCARE-VEDSv20-2004.pdf (2004)
28. Thomson, M., Winterbottom, J., et. al.: Representation of Uncertainty and Confidence in the Presence Information Data Format Location Object (PIDF-LO), RFC7459, The Internet Engineering Task Force. https://tools.ietf.org/html/rfc7459 (2015)
29. Intelligent Transport Systems – eSafety – eCall Minimum Set of Data, I.S. EN 15722:2015, Irish Standard, 22 Apr 2015
30. Winterbottom, J., et. al.: GEOPRIV Presence Information Data Format Location Object (PIDF-LO) Usage Clarification, Considerations, and Recommendations, RFC5491, The Internet Engineering Task Force. https://tools.ietf.org/html/rfc5491 (2009)
31. Gellens, R., et.al.: Additional Data Related to an Emergency Call, RFC7852, The Internet Engineering Task Force. https://tools.ietf.org/html/rfc7852 (2016)
32. Rosen, B., et.al.: Framework for Emergency Calling Using Internet Multimedia, RFC6443, The Internet Engineering Task Force. https://tools.ietf.org/html/rfc6443 (2011)
33. Rosen, B., Polk, J.: Best Current Practice for Communications Services in Support of Emergency Calling, RFC6881, The Internet Engineering Task Force. https://tools.ietf.org/html/rfc6881 (2013)
34. Schulzrinne, H., et al.: Dynamic Extensions to the Presence Information Data Format Location Object (PIDF-LO), RFC 5962, The Internet Engineering Task Force (2010)

19. Winterbottom, A., et al.: Pan-European Mobile Emergency Applications (PEMEA) Requirements and Functional Architecture, EENA NG112 Technical Committee Document. European Emergency Number Association. http://www.eena.org/download.asp?item_id=158 (2015)
20. Gellens, R., et al.: Additional Data Related to an Emergency Call. The Internet Engineering Task Force. https://tools.ietf.org/html/draft-ietf-ecrit-additional-data-38 (2016)
21. Winterbottom, J. et al.: Specifying Civic Address Extensions in the Presence Information Data Format Location Object (PIDF-LO). [illegible]

[illegible]

Towards Multi-layer Interoperability of Heterogeneous IoT Platforms: The INTER-IoT Approach

Giancarlo Fortino, Claudio Savaglio, Carlos E. Palau, Jara Suarez de Puga, Maria Ganzha, Marcin Paprzycki, Miguel Montesinos, Antonio Liotta and Miguel Llop

Abstract Open interoperability delivers on the promise of enabling vendors and developers to interact and interoperate, without interfering with anyone's ability to compete by delivering a superior product and experience. In the absence of global IoT standards, the INTER-IoT voluntary approach will support and make it easy for any

---

The original version of this chapter was revised: Incorrect author name has been corrected. The erratum to the book is available at https://doi.org/10.1007/978-3-319-61300-0_11

G. Fortino (✉) · C. Savaglio
DIMES - University of Calabria, Via P. Bucci, 87036 Rende (CS), Italy
e-mail: g.fortino@unical.it

C. Savaglio
e-mail: csavaglio@dimes.unical.it

C.E. Palau · J.S. de Puga
DCOM – Universitat Politecnica de Valencia, Camino de Vera S/N, 46022 Valencia, Spain
e-mail: cpalau@dcom.upv.es

J.S. de Puga
e-mail: jasuade@dcom.upv.es

M. Ganzha · M. Paprzycki
Systems Research Institute, Polish Academy of Sciences, ul. Newelska 6, 01-447 Warsaw, Poland
e-mail: mganzha@dcom.upv.es

M. Paprzycki
e-mail: marcin@dcom.upv.es

M. Montesinos
PRODEVELOP, PlaÇa de Joan de Vilarasa, 14-5, 46001 Valencia, Spain
e-mail: mmontesinos@prodevelop.es

A. Liotta
Technische Universiteit Eindhoven, P.O. Box 513, 5600 MB Eindhoven, The Netherlands
e-mail: aliotta@tue.nl

M. Llop
Avinguda Moll del Turia s/n, 46024 Valencia, Spain
e-mail: MLlop@fundacion.valenciaport.com

R. Gravina et al. (eds.), *Integration, Interconnection, and Interoperability of IoT Systems*, Internet of Things, DOI 10.1007/978-3-319-61300-0_10

IoT stakeholder to design open IoT devices, smart objects, services, and complex systems and get them to be operative and interconnected quickly, thus creating new IoT interoperable ecosystems by using a bottom-up approach. In particular, INTER-IoT is based on hardware/software tools (INTER-Layer) granting multi-layer interoperability among IoT system layers (i.e. device, networking, middleware, application service, data and semantics), on frameworks for open IoT application and system programming and deployment (INTER-FW), and on a full-fledged CASE tool-supported engineering methodology for IoT systems integration (INTER-Meth). The INTER-IoT approach is notably exemplified through two use cases: INTER-LogP, involving interoperability of port logistics ecosystems, and INTER-Health, encompassing integration between e-Health at home and in mobility infrastructures.

Keywords Internet of things · Interoperability · Platforms · Device · Networking · Middleware · Application services · Data · Semantics · e-Health · Smart port logistics

1 Introduction

In recent years, due to a great interest of both Industry and Academy in researching and developing Internet of Things (IoT) technology [23, 44], many solutions at different levels (from the IoT device-level to full-fledged IoT platforms) have been implemented. However, there is no well-established reference standard for IoT platform technology and we do not foresee one in the near future. Hence, IoT scenarios will be characterized by a high-degree of heterogeneity at all levels (device, networking, middleware, application service, data/semantics), preventing interoperability of IoT solutions [10, 28].

Lack of interoperability causes major technological and business issues such as impossibility to plug non-interoperable IoT devices into heterogeneous IoT platforms, impossibility to develop IoT applications exploiting multiple platforms in homogeneous and/or cross domains, slowness of IoT technology introduction at a large-scale, discouragement in adopting IoT technology, increase of costs, scarce reusability of technical solutions, and user dissatisfaction [22, 40].

A multi-layered approach to integrate heterogeneous IoT devices, networks, platforms, services and data will allow heterogeneous elements to cooperate seamlessly to share information, infrastructures and services as in a homogenous scenario [1, 49]. Thus, the main goal of the INTER-IoT approach being developed in the EU-funded H2020 INTER-IoT project is to comprehensively address the lack of interoperability in the IoT realm by proposing a full-fledged approach facilitating "voluntary interoperability" at any level of IoT platforms and across any IoT application domain, thus guaranteeing a seamless integration of heterogeneous IoT technology.[1] The proposed approach will allow effective and efficient development of adaptive, smart IoT applications and services atop different heterogeneous IoT platforms, spanning single and/or multiple application domains. INTER-IoT focuses in two application domains.

[1]INTER-IoT project website: http://www.inter-iot.eu.

The INTER-IoT approach thus aims to provide open interoperability, which delivers on the capability of enabling vendors and developers to interact and interoperate, without interfering with anyone's ability to compete by delivering a superior product and experience. In the absence of global IoT standards, the INTER-IoT project is supporting and making it easy for any company to design IoT devices, smart objects, and/or services and get them to the market quickly, thus creating new IoT interoperable ecosystems.

This chapter is organized as follows. In Sect. 2, a state-of-the-art analysis on IoT platform interoperability approaches is presented. Section 3 describes the INTER-IoT approach, specifically detailing the technical solutions defined for IoT interoperability. In Sect. 4, we introduce the two main use cases that will be developed in the project. Finally, conclusions are drawn and future work is delineated.

2 Related Work

Interoperability among heterogeneous systems can be understood and involves [27]:

- *Technical Interoperability*, which is associated with hardware/software components, systems and platforms that enable machine-to-machine communication to take place. This kind of interoperability is often centered on (communication) protocols and the infrastructure needed for those protocols to operate.
- *Syntactical Interoperability*, which is associated with data formats.
- *Semantic Interoperability*, which is associated with the meaning of content and concerns the human rather than machine interpretation of the content.
- *Organizational Interoperability*, which is the ability of organizations to effectively communicate and transfer (meaningful) data (information) across different information systems over widely different infrastructures. Organizational interoperability depends on the former three.

In the following subsections, we briefly overview platforms and projects (Sects. 2.1 and 2.2) and standardization efforts (Sect. 2.3) strongly correlated to IoT systems interoperability and also highlight their links to INTER-IoT.

2.1 Projects Related with IoT and IoT Platforms

Several projects funded in previous years by the European Commission and other international organizations worldwide have been focusing on domain-specific and/or open IoT platforms. In Table 1, the most known and diffused ones are reported. In particular, a short description is provided along with similarities and differences with respect to our INTER-IoT approach. Considering that INTER-IoT is not presenting a new IoT platform but an interoperability framework.

Table 1 Representative projects proposing general-purpose and domain-specific IoT architectures/platforms

International/European R & D & I activities	Link to INTER-IoT
IOT-A[a] Creation of an architectural reference model together with the definition of an initial set of key building blocks for enabling the emerging IoT	*Similarities*: Reference architecture for IoT platforms. *Differences*: No development of a concrete platform for interoperability among other IoT platforms. No integration methodology is provided
COMPOSE[b] Creation of an ecosystem transforming the IoT into an Internet of Services, through an open marketplace	*Similarities*: Promotion of a bottom-up approach for IoT ecosystems development. *Differences*: Not specifically addressing interoperability issues. No integration methodology is provided
Web of Objects[c] Development of a network and services infrastructure for autonomic cooperating smart objects	*Similarities*: Reference architecture and virtualization as means for integration. *Differences*: Not focused on granting interoperability through integration. No methodology is provided
iCore[d] Development of a cognitive management framework for the reliable mash-up of smart objects and smart services	*Similarities*: Interoperability between IoT devices and IoT services through virtualization. *Differences*: Global and layer-oriented interoperability is not addressed. No integration methodology is provided
Butler[e] Integration of current IoT technology and development of new technologies to form a "bundle" of applications, platform features and services, emphasising pervasiveness, context-awareness and security for IoT	*Similarities*: Interoperability at device-level through Smart Gateway. *Differences*: No global and layer-based interoperability is provided and neither an integration methodology
IoTLaB[f] Creation of a crowdsourcing infrastructure together with the supporting mechanisms that will enable multidisciplinary experimentation platform in the general domain of IoT	*Similarities*: Resource virtualization for virtual interconnection of networks and devices. *Differences*: No global and layer-based interoperability is provided and neither an integration methodology
IoT@Work[g] Development of an IoT-based plug and work concept centered on industrial automation, specifically enabling IoT applications in automation domains	*Similarities*: Network-oriented interoperability among industrial IoT devices. *Differences*: Special-purpose domain (industrial plants). No global and layer-based interoperability (apart from the network-layer) is provided and neither an integration methodology
OpenIoT[h] Development of an open source middleware for collecting information from sensor clouds of heterogeneous domains and offering utility-based IoT services	*Similarities*: Interoperability at device level through Global Sensor Network. *Differences*: No platform integration infrastructures and methodology available. Data-level integration of heterogeneous IoT sources

(continued)

Table 1 (continued)

International/European R & D & I activities	Link to INTER-IoT
CASAGRAS/CASAGRAS2[i] Provision of a framework of foundation studies concerning radio frequency identification (RFID) with particular reference to the emerging IoT	*Similarities*: Interoperability between RFID-oriented devices through an Interrogator-Gateway Layer. *Differences*: Strongly RFID oriented. No applicability in more general IoT contexts. No platform integration infrastructures and methodology available
Smart Santander[j] Development of city-scale experimental research facility to support typical applications and services for smart cities, exploiting existing frameworks as WISEBED, SENSEI and USN	*Similarities*: Integration approach of exploiting other framework/platform strengths. *Differences*: Heterogeneity across three specific frameworks but not across layers; no methodology or reference architectures provided; very focused on the Smart City domain
TRESCIMO[k] Aims at developing a M2M platform that links Smart Applications to Sensor Networks and low level devices w.r.t. smart energy management and smart grid	*Similarities*: M2M-based reference architecture. *Differences*: Single domain (Smart Cities) and specific focus on M2M and energy. Thus, global and layer-based interoperability is not provided; no integration methodology is proposed
SunRise[l] Creation of a federation of experimental facilities covering the diverse marine environments allowing researchers to experiment with novel paradigms for the Internet of Underwater Things	*Similarities*: Efforts in defining standard interoperable methods and architectures. *Differences*: Focus on underwater IoT and its peculiar communication issues. No methodology for IoT platform interoperability
Webinos[m] Development of an Open Source Platform and software components for the FI to enable web applications and services to be used and shared consistently and securely over converged and connected devices	*Similarities*: Provision of interoperability and interconnectivity by means of gateway and open Web standards. *Differences*: No full-fledged layer-oriented interoperability. Focus is on software concepts based on web-based services
SANDS[n] Development of a physical and computational networked infrastructure for household appliances, forming an IoT ecosystem, to meet the needs of their owners	*Similarities*: Open (social-oriented) IoT architecture. *Differences*: Specific focus on domotic. Not addressing platform interoperability and integration methodology
VITAL[o]: Development of a Cloud-of-Things-based platform integrating and interacting with a multitude of different IoT data sources and systems within the Smart City domain	*Similarities*: Lightweight open meta-architecture for IoT frameworks integration based on voluntary interoperability. *Differences*: The integration meta-architecture provides global integration and not layered-oriented integration that would allow higher performance and reduce reliability and security issues. Not providing any systematic methodology for IoT platforms integration

(continued)

Table 1 (continued)

International/European R & D & I activities	Link to INTER-IoT
FIWARE[p] development of a middleware for the Future Internet based on Generic and Specific enablers, it includes several components to integrate IoT and services	*Similarities*: Approach based on the deployment of enablers and integration of services. *Differences*: Device and network layers are transparent to the framework of IoT

[a]Internet Of Things – Architecture, available at http://www.iot-a.eu
[b]Collaborative Open Market to Place Objects at your Service, available at http://www.compose-project.eu
[c]Web of Objects, available at http://www.web-of-objects.com/
[d]Internet Connected Objects for Reconfigurable Eco-systems, available at http://www.iot-icore.eu/
[e]uBiquitous, secUre inTernet-of-things with Location and contEx-awaReness, available at http://www.iot-butler.eu/
[f]Internet-of-Things Laboratory, available at http://www.iotlab.eu
[g]Internet-of-Things at Work, available at http://www.iot-at-work.eu/
[h]Open Internet-of-Things, available at http://www.openiot.eu
[i]Coordination and Support Action for Global RFID-related Activities and Standardisation, available at http://www.iot-casagras.org
[j]Smart Santander Project, available at http://www.smartsantander.eu/
[k]Testbeds for Reliable Smart City Machine to Machine Communication, available at http://trescimo.eu/
[l]Sensing, monitoring and actuating on the UNderwater world through a federated Research InfraStructure Extending the Future Internet, available at http://fp7-sunrise.eu
[m]Secure WebOS Application Delivery Environment, available at http://webinos.org
[n]Social AND Smart, available at http://www.sands-project.eu/
[o]Virtualized programmable InTerfAces for smart, secure and cost-effective IoT depLoyments in smart cities, available at http://vital-iot.eu
[p]FI-PPP Future Internet Core Platform, available at http://www.fiware.org

2.2 *IoT-EPI*

IoT-EPI is a European Initiative addressing the new EU-funded H2020 programs about IoT platform and interoperability development. At the core of IoT-EPI are the following seven research and innovation projects: INTER-IoT (i.e., the subject of this chapter), BIG IoT, AGILE, symbIoTe, TagItSmart!, VICINITY and bIoTope. The European Platforms Initiative is coordinated by two Collaborative Support Actions (CSAs): Unify-IoT and Be-IoT. In Table 2 a brief description of the projects along with similarities and differences with respect to INTER-IoT is reported.

2.3 *IoT Standardization*

Currently several standardization efforts are underway to define architectural standards for IoT systems interoperability. The most important ones are reported in Table 3 along with a comparison with the INTER-IoT approach. It worth noting that the main difference is that INTER-IoT aims at voluntary (i.e., non standards-oriented) interoperability.

Table 2 IoT-EPI projects

Project	Link to INTER-IoT
BIG IoT[a] aims at establish interoperability by defining a unified Web API for IoT platforms: the BIG IoT API. This Web API is aligned with the standards currently developed by the W3C Web of Things group. An IoT platform or service implements the API to register and access the BIG IoT Marketplace so as to interoperate with services/applications available in the marketplace. BIG IoT pilots involve Smart City applications	*Similarities*: it is not developing yet another IoT platform but the aim is higher-level interoperability of already existing heterogeneous IoT services and applications. It is worth noting that the BIG IoT API has a similar role of the INTER-FW API. *Differences*: the approach only offers a high-level API to grant application and/or service interoperability and does not provide methods and methodology to integrate IoT platforms at the different finer-grain layers identified by INTER-IoT. Moreover, the use cases are in a different application domain
AGILE[b] (An Adaptive and Modular Gateway for the IoT) builds a modular and adaptive gateway for IoT devices. Modularity at the hardware level provides support for various wireless and wired IoT networking technologies (KNX, ZWave, ZigBee, BLE, etc.). At the software level, different components enable new features: data collection and management on the gateway, intuitive interface for device management, visual workflow editor for creating IoT apps, and an IoT marketplace for installing IoT apps locally. AGILE pilots involve open field and animal monitoring, enhanced retail services, people monitoring based on wearables	*Similarities*: the device layer provides interoperability among heterogeneous devices based on different communication protocols. The use case based on wearables is similar somehow to the INTER-Health use case. *Differences*: the approach only offers device layer interoperability and does not provide methods and methodology to integrate IoT platforms at the different finer-grain layers identified by INTER-IoT. Moreover, the other use cases are in a different application domain
symbIoTe[c] (symbiosis of smart objects across IoT environments) will enable the discovery and sharing of resources for rapid cross-platform application development and will facilitate the blending of next generation of smart objects with surrounding environments. symbIoTe will achieve all of the above by designing and implementing an Open Source mediation prototype. Its pilots encompass several smart environments and smart mobility	*Similarities*: it steps into the IoT landscape to devise an interoperability framework across existing and future IoT platforms. Specifically, like INTER-IoT it chooses the challenging task to implement IoT platform federations so that they can securely interoperate, collaborate and share resources for the mutual benefit, also supporting the migration of smart objects between various IoT domains and platforms. *Differences*: the approach is basically based on a mediation prototype to support interoperability and does not provide methods and methodology to integrate IoT platforms at the different finer-grain layers identified by INTER-IoT. Moreover, the use cases are in different application domains

(continued)

Table 2 (continued)

Project	Link to INTER-IoT
TagItSmart![d] (Smart Tags driven service platform for enabling ecosystems of connected objects) has the objective of creating a set of tools and enabling technologies integrated into a platform with open interfaces enabling users across the value chain to fully exploit the power of condition-dependent FunCodes to connect mass-market products with the digital world across multiple application sectors. Its pilots are related to from preproduction to recycling smart chains	*Similarities*: the aim to define open interface for easing interconnection, even though interoperability is not a main issue to deal with. The project proposes the creation of a developers community which will be providing new services in a market place directly using the API and the funny tags developed in the project. *Differences*: the approach is not aimed at interoperability of heterogeneous IoT systems based on integration/interconnection methods and methodologies. Moreover, the use cases are in different application domains
VICINITY[e] (Open virtual neighbourhood network to connect IoT infrastructures and smart objects) aims to provide the owners of connected IoT infrastructures with a decentralized interoperability. It connects different smart objects into a "social network" called virtual neighbourhood where infrastructure owners keep under control their shared devices and data thanks to web based operator console called VICINITY neighbourhood manager (VNM). Guest IoT infrastructures, VICINITY enabled services as well as the VICINITY auto-discovery space are connected to a VICINITY interoperability gateway using the same VICINITY gateway API. Use cases are in the energy, building, e-Health, and mobility application domains	*Similarities*: it grants decentralized interoperability through interconnection of heterogeneous systems through gateways and data concentrators. The project proposes the definition and development of an ontology to allow semantic interoperability between IoT platforms (cloud and gateway level) related with the associated use cases. *Differences*: the approach is basically (standard or proprietary) based on infrastructure gateways to support interoperability but does not provide methods and methodology to integrate IoT platforms at the different finer-grain layers identified by INTER-IoT. Moreover, the use cases are in different application domains apart from the e-Health use case
bIoTope[f] (Building an IoT OPen innovation Ecosystem for connected smart objects) provides the necessary standardised Open APIs to enable the publication, consumption and composition of heterogeneous information sources and services from across various platforms, including FI-WARE, OpenIoT, city dashboards, etc. Pilots are in the area of Smart Cities	*Similarities*: interconnection of heterogeneous IoT platforms and systems through common API, the project looks for specific standards to achieve interoperability. *Differences*: the approach only offers a high-level API to grant systems of systems interconnection and does not provide methods and methodology to integrate IoT platforms at the different finer-grain layers identified by INTER-IoT. Moreover, the use cases are in a different application domain

[a]BIG IoT - Bridging the Interoperability gap of the Internet of Things, http://big-iot.eu/
[b]AGILE - An Adaptive and Modular Gateway for the IoT, http://agile.eu/
[c]symbIoTe - symbiosis of smart objects across IoT environments, https://www.symbiote-h2020.eu/
[d]TagItSmart! - Smart Tags driven service platform for enabling ecosystems of connected objects, http://www.tagitsmart.eu/
[e]VICINITY - Open virtual neigh-bourhood network to connect IoT infra-structures and smart objects, http://vicinity2020.eu/vicinity/
[f]bIoTope - Building an IoT OPen innovation Ecosystem for connected smart objects, http://biotope.cs.hut.fi/

Table 3 IoT Standardization Initiatives

Standardization initiative	Link to INTER-IoT
AIOTI[a] was initiated by the European Commission in 2015, with the aim to strengthen the dialogue and interaction among IoT players in Europe, and to contribute to the creation of a dynamic European IoT ecosystem to speed up the take up of IoT	*Similarities*: AIOTI uses the same architecture reference model (ARM) for IoT exploited by INTER-IoT. Such ARM derives from the IoT-A project. *Differences*: the aim of INTER-IoT is not to propose a standard but to interconnect heterogeneous systems based on even different standards or proprietary solutions
IEEE P2413[b] is a standard that defines an architectural framework for the IoT, including descriptions of various IoT domains, definitions of IoT domain abstractions, and identification of commonalities between different IoT domains. The architectural framework for IoT provides a reference model that defines relationships among various IoT verticals (e.g., transportation, healthcare, etc.) and common architecture elements. It also provides a blueprint for data abstraction and the quality "quadruple" trust that includes protection, security, privacy, and safety	*Similarities*: IEEE P2413 is based on an ARM that is similar to the IOT-A ARM on which INTER-IoT is based. Moreover, both reference models share several commonalities that are used to extend INTER-IoT Interoperability Reference Model. *Differences*: the aim of INTER-IoT is not to propose a standard but to interconnect heterogeneous systems based on even different standards or proprietary solutions. The reference model provided by INTER-IoT does not have the aim to be used as a reference for developing new open platforms, but for allow interoperability of existing platforms
oneM2M[c] has the purpose and goal of developing technical specifications which address the need for a common M2M Service Layer that can be readily embedded within various hardware and software, and relied upon to connect the myriad of devices in the field with M2M and IoT application servers worldwide	*Similarities*: a Service Layer granting access to heterogeneous machines and subsystems. Additionally, INTER-IoT considers gateway nodes in order to allow device to device and network interoperability. *Differences*: INTER-IoT does not aim at defining standard specifications but is based on voluntary interoperability concept. However, the use of standards may allow a broader connectivity with other nodes

[a]The Alliance for Internet of Things Innovation (AIOTI), http://www.aioti.org/
[b]P2413 - Standard for an Architectural Framework for the Internet of Things (IoT), https://standards.ieee.org/develop/project/2413.html
[c]OneM2M - Standards for M2M and the Internet of Things, http://www.onem2m.org/

3 The INTER-IoT Approach

The solution adopted by INTER-IoT includes three main solutions to grant voluntary interoperability (see Fig. 1):

- *INTER-LAYER*: methods and tools for providing interoperability among and across each layer (virtual gateways/devices, network, middleware, application services, data and semantics) of IoT platforms. Specifically, we will explore real/virtual

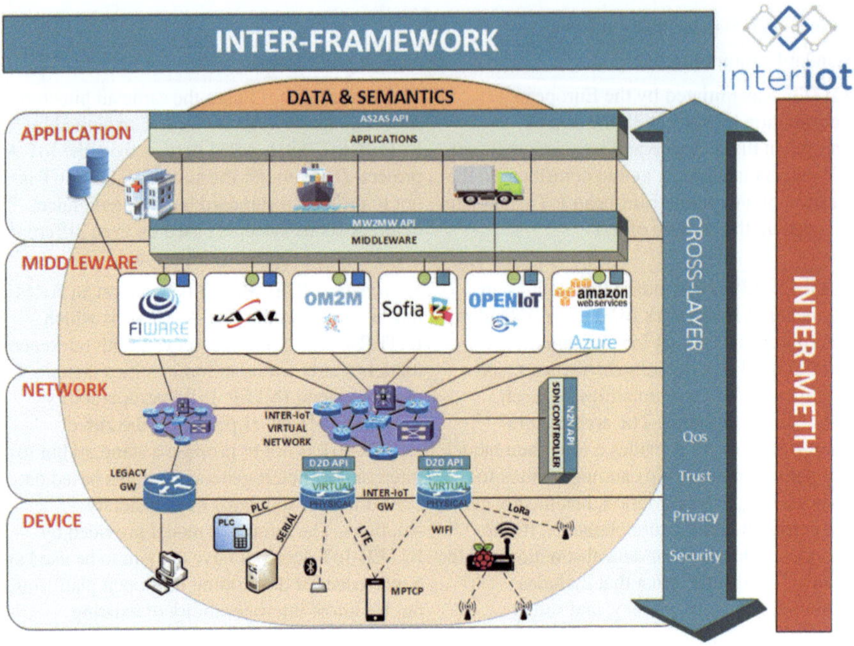

Fig. 1 The INTER-IoT abstract architecture highlighting INTER-IoT multi-layer solutions (device, networking, middleware, application, data and semantics), their interconnection (cross-layer), and their tools (INTER-FrameWork and INTER-METH)

gateways [2, 36] for device-to-device communication, virtual switches based on SDN for network-to-network interconnection, super middleware for middleware-to-middleware integration, service broker for the orchestration of the service layer and a semantics mediator for data and semantics interoperability [1, 20, 21].

- *INTER-FW*: a global framework (based on an interoperable meta-architecture and meta-data model) for programming and managing interoperable IoT platforms, including an API to access INTER-LAYER components and allow the creation of an ecosystem of IoT applications and services. INTER-FW will provide management functions specifically devoted to the interconnection between layers. The provided API includes security and privacy features and will support the creation of a community of users and developers.
- *INTER-METH*: an engineering methodology based on CASE (Computer Aided Software Engineering) tool for systematically driving the integration/interconnection of heterogeneous non-interoperable IoT platforms.

3.1 INTER-LAYER

Differently from current interoperability approaches (see Tables 1 and 2), INTER-IoT uses a layer-oriented approach to fully exploit specific functionalities of each

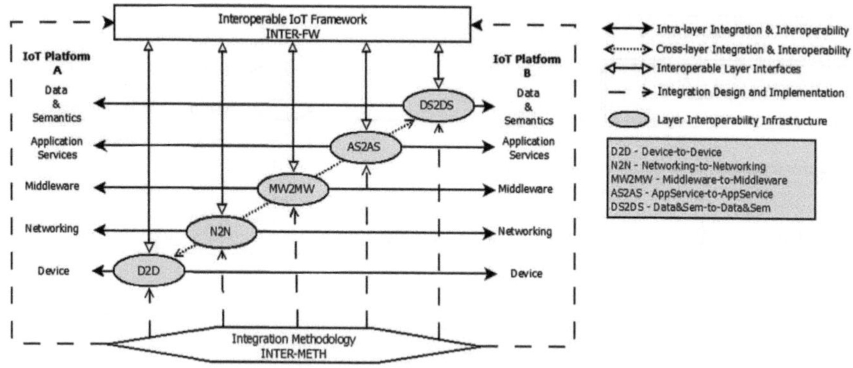

Fig. 2 Abstract schema of INTER-LAYER

layer (device, networking, middleware, application services, data and semantics). Although the development of a layer-oriented approach is a research challenge, as compared to an application-level approach, it has a higher potential to deliver tight bidirectional integration among heterogeneous IoT platforms, notably guaranteeing independence, thus providing higher performance, modularity, adaptability, flexibility, reliability, security, privacy and trust. Furthermore, what is extremely important, INTER-IoT will give more control on functional and non-functional requirements.

As highlighted in Fig. 2, INTER-IoT approach is based on the following real/virtual layer interoperability infrastructures among peer layers guaranteeing interoperability/integration: Device-to-Device (D2D); Networking-to-Networking (N2N); Middleware-to-Middleware (MW2MW); Application Services-to-Application Services (AS2AS); Data and Semantics-to-Data and Semantics (DS2DS). We will investigate two main types of D2Ds: smart device gateways and device virtualization wrappers. N2N will be based on Network Functions Virtualization components, representing the gateways adapted to different protocols; also the virtual gateways will be connected using Software Defined Network enabled switches. The use of NFV and SDN through a software controller provides extra flexibility and adequate management of data flows priorities and QoS. MW2MW will rely on smart brokers coordinating between heterogeneous middleware manager components. Virtualization will be exploited to develop AS2AS due to the effectiveness and flexibility that service virtualization can provide at application level. Finally, DS2DS will be designed around smart data and semantics management concepts. Every interoperability mechanism will be accesses through INTER-API of INTER-FW, see Sect. 3.2.

Each layer interoperability infrastructure (implementable in hardware, software, or both) not only provides strong coupling between peer layers but also exposes an interface, which can be programmed to control/interact with the component. Interfaces will be controlled by a meta-level framework to provide global interoperability (see Fig. 3). Moreover, the layer interoperability infrastructures can communicate with each other to provide cross-layering that aims at strengthening integration among layers so providing more efficiency and reliability, while still supporting flexibility

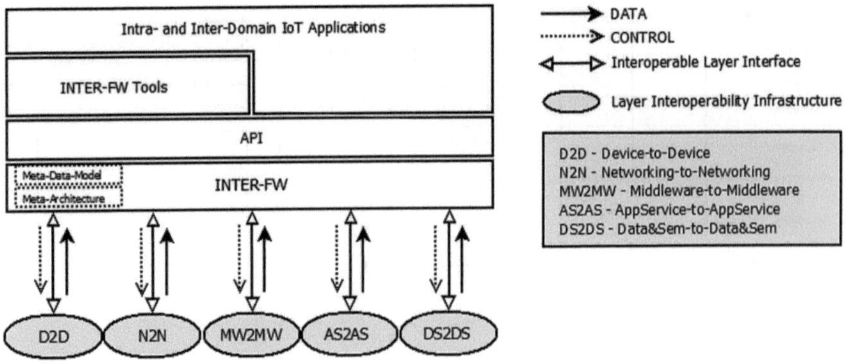

Fig. 3 Abstract schema of INTER-FW

and adaptability. Cross-layer component is fully devoted to the support and coordination of security and privacy mechanisms and services for the whole INTER-Layer.

3.1.1 Device

As sensors, actuators and smart devices become smaller, more versatile, lower cost and more power efficient, they are being deployed in greater numbers, either as special-purpose devices or embedded into other products. The unification and convergence of the vast number of platforms already deployed, the accessibility (API and interfaces) of the platform to app developers, requires interoperability. Smart-phones are key components in Device-to-Device (D2D) communication and interoperability, however there are many other types of devices that are currently deployed, both independently (e.g., smart watches and other wearables) and as part of other devices and platforms (e.g., consumer electronics or Cyber-Physical Systems).

Different communication protocols are used at device level. Here, Cellular and WiFi that are ubiquitous; they are evolving to support higher bandwidths and lower cost. Bluetooth is also becoming lower cost. New communication technologies like Bluetooth low energy (Bluetooth LE) and NFC are opening new possibilities for IoT. However, also traditional communication protocols and mechanisms for sensors, actuators and smart objects have to be considered (e.g., ZigBee, ISA100, WirelessHart, LoRa or SigFox), in addition to other non-standard proprietary protocols developed by individual vendors.

Different classes of IoT objects need different communication supports: e.g. 'deterministic' communication protocols (MAC and Routing layers) are not possible using current Internet protocols, but may be needed by some application. Standardization on these topics is just starting (e.g., detnet working group in IETF). Yet, deterministic communications will hardly meet the interoperability requirements of all IoT objects. Typically, device-level interconnection of IoT architectures has been performed using gateway-based solutions. FP7 Butler project (Table 2) proposed a device-centric architecture where a SmartGateway allows interconnection between smart objects (sensors, actuators, gateways) using IPv6 as communication protocol. Different approaches have been developed to integrate and interoperate devices in IoT architectures. Basic devices (e.g., sensors, tags, actuators) are virtualized and can be composed in more complex smart systems [4]. The idea has been to create virtual objects, allowing object composition, considering a virtual object as a counterpart of existing smart objects [7].

INTER-IoT will provide fundamental benefits and competitiveness improvements in the way IoT devices will communicate with each other and will interface with different IoT platforms and subsystems. One of the proposed progresses regarding D2D interaction is to complement standardized communication protocols (which are mostly deterministic and reactive) with an ability for objects to make sense of their surroundings in order to understand how to best interplay with their neighbours. This requires new 'proactive' and 'predictive' communications capabilities, whereby a node can determine its communication requirements and those of its neighbours well before communication is required [29]. It has recently been proven that machine learning capabilities can run even on small sensors (with as little as 20 Kbytes of RAM) [6]. **INTER-IoT** is developing an interoperable communication layer that accommodate for opportunistic communications among heterogeneous nodes/devices, based on prediction mechanisms.

In particular, interoperability at device level implies that (i) heterogeneous IoT devices are able to interact with each other; (ii) heterogeneous IoT devices can be accessed through a unifying interface although they use different communication protocols; heterogeneous IoT devices can be integrated into any IoT platform. Regarding interoperability, we are exploring two different approaches: gateway-based and virtualization. In the former, a gateway-based approach are defined to adapt different communication protocols/languages on which heterogeneous devices are based. In the latter, virtualization techniques are defined to create virtual IoT devices that can be accessed through a unified service-oriented interface and interact through application-level interaction protocols.

With regards to integration, we are reusing the approaches for interoperability and further exploit wrapping methods to integrate IoT devices into non-interoperable IoT platform. We also consider fully reconfigurable devices supporting the emerging SDR (Software Defined Radio) paradigm. In particular, these specific devices can be used to implement communication interoperability by acting as "transparent bridges or gateways" between different radio technologies. This approach could be very effective when different smart objects, placed in a specific environment and equipped with different low range radio technologies such as Bluetooth, ZigBee, Wi-Fi, would communicate with each other. The deployment of novel SDR hardware and software architectures can effectively help in solving many internetworking issues at lower levels of the architecture. Coupling heterogeneous devices and/or integrating them into heterogeneous IoT platforms could bring to reliability (in broad sense), security and trust issues that need to be addressed to have fully interoperability at functional and non-functional level. INTER-IoT is analyzing in depth such issues to provide techniques with the required level of reliability, security and trust, and to be compliant with the different recommendations and guidelines from security experts and standardization organizations.

3.1.2 Networking

IoT products will encompass different data communication scenarios. Some may involve sensors that send small data packets infrequently and do not prioritize timely delivery. Others may involve storage in order to sustain periods when the communication link is down (e.g., Delay Tolerant Networks). Others may need high bandwidth but be able to accept high latency. And others may need high quality, high bandwidth, and low latency. Large amounts of traffic with relatively short packet sizes will require sophisticated traffic management and traffic engineering procedures. More efficient protocols and management mechanisms will help reducing overheads but may present challenges to system integrity, reliability and scalability. Interface standardization is desirable so that IoT objects can communicate quickly and efficiently, and allow mobility between interoperable IoT platforms. IoT objects will need a way to quickly and easily discover each other and learn their neighbour's capabilities.

At networking layer different protocols can be used like 6LowPAN, TCP/HTTP, UDP/CoAP. Communication between real objects and the gateway can be based on universal plug and play (UPnP) or DLNA. Use of buses based on MQTT protocol can also be used to implement asynchronous communications between entities. The most promoted networking protocol in IoT environments is IPv6 and its version for constrained devices 6LoWPAN, even though its adoption is slow, and without global adoption it will be impossible for IoT to proliferate. IPv6 provides the following benefits to IoT configurations: (i) IPv6 auto-configuration; (ii) Scalable address space (sufficiently large for the enormous numbers of IoT objects envisioned); (iii) Redefined headers that enable header compression formats; (iv) Easy control of the system of things; (v) Open/Standard communications; (vi) IPv6 to IPv4 transition methods; (vii) IPv6 over constrained node networks (6LO, 6LoWPAN).

IoT platforms have usually mechanisms for integrating with external systems, but they are all based on specific point-to-point connections, usually with legacy systems in the area of interest of the IoT platform (e.g. city, neighbourhood, factory, hospital, port, house, etc.). The integration between IoT platforms will allow tracking the behaviour of these objects when they move outside the intrinsic area of interest and get into the area of interest of another IoT platform. The pub/sub mechanism usually available in the communication buses at the core of these IoT platforms and the possible object context sharing allow a powerful and easy way to track the behaviour of these objects among different IoTs scope areas.

INTER-IoT will provide support for as many networks as possible. Main contributions of the project are focused on multi-homing capabilities among the different IoT objects in order to provide network offloading connectivity and seamless mobility between different IoT platforms of moving objects. **INTER-IoT** is using SDN components to configure interconnection at network level(including an OpenFlow software controller) as support for interoperability and roaming of smart objects between different platforms of the same ecosystem while keeping secure connectivity and also guaranteed quality of service. Resource management and scalability so as reliability, trust, privacy and security are non-functional requirements that will be addressed.

In particular, **INTER-IoT** is defining and analysing methods to integrate and allow interoperability of IoT components at network layer. It will specifically address: (i) pervasiveness and ubiquity network aspects including seamless mobility of smart objects between different IoT architectures in both use cases; (ii) the highly constrained environment in terms of physical size, available memory, CPU power and battery life in addition to communicating over wireless low power lossy networks in which operate Smart Objects; (iii) research in routing mechanisms to overcome traditional routing protocols drawbacks, specially route of information over IPv6, in particular, carrying IPv6 over low power networks (6LowPAN) and RPL (Routing Protocol for Low Power and Lossy Networks). Another key issue to be developed in the task will be IoT device mobility within different IoT architectures, including (i) network offloading with connectivity to different access networks; (ii) multihoming and (iii) secure seamless mobility.

3.1.3 Middleware

Middleware, widely used in conventional distributed systems [5], is a fundamental tool for the design and implementation of both IoT devices and IoT systems [23]. They provide general and specific abstractions (e.g., object computation model, inter-object communication, sensory/actuation interfaces, discovery service, knowledge management), as well as development and deployment tools, through which IoT devices, IoT systems and their related applications can be easily built up. Indeed, middleware (i) enable connectivity for huge numbers of diverse components comprised at Device Layer, (ii) realize their seamless interoperability at Networking Layer, and (iii) ensure operational transparency at Application Service Layer. In such a way, heterogeneous, often complex and already existing IoT devices and IoT systems, belonging to different application domains and not originally designed to be connected, can be easily integrated, effectively managed and jointly exploited. It follows that the role of middleware within the cyberphysical, heterogeneous, large scale and interconnected IoT scenario is even more crucial than within conventional distributed systems.

Over the years, many IoT middleware have been proposed, so much so that only in [3, 14, 37] more than 70 contributions have been surveyed and compared. The best way to analyse such plethora of middleware, regardless of the specific detail or technology, is to build up comparison frameworks around well-defined criteria to effectively highlight their salient differences and similarities. In such direction, middleware have been compared:

- in [14], according to their requirements at device or system levels;

- in [37], according to their functional, not functional and architectural requirements, as well as to their design approaches (application-specific, service oriented, agent-based, etc.);
- in [3], according to their functional requirements and supported low-level interface protocols.

Taking into account these three contributes and their related comparison frameworks, in Table 4 we identified eight recurrent criteria, which can be hence considered as IoT middleware main features, and the four IoT middleware that better fulfil them. In very few words, LinkSmart [8] is service-based middleware for ambient intelligence (AmI) systems, supporting devices communication, virtualization, dynamic reconfiguration, self-configuration, energy optimization and security by means of WebService-based mechanisms enriched by semantic resolution. UbiROAD [42] is sematic, context-aware, self-adaptive agent-based middleware for smart road environments, aiming at collecting, analysing and mining real time data from in-car and roadside heterogeneous devices. ACOSO [11, 12, 40] is an agent-oriented middleware with a related methodology [13] fully supporting the development (from the modelling to the implementation phase [18, 19]), management and deployment of smart objects and IoT systems, as well as their integration with the Cloud [15, 17]. IMPReSS [26], finally, is a middleware conceived for the rapid development of context-aware, adaptive and real-time monitoring applications to control and optimize energy usage in smart cities. Table 4 shows how these four middleware (Totally, Partially or Not) fulfil the eight IoT middleware main features.

In particular, **INTER-IoT** will focus on defining component-based methods for middleware interoperability/integration; in particular, we focus on discovery, management and high-level communication of IoT devices in heterogeneous IoT platforms. We will define two main approaches: (i) definition of overlay middleware components able to couple the middleware components of the heterogeneous IoT platforms; (ii) virtualization of the heterogeneous middleware components. In the first approach, we will design overlay middle components such as mediators and brokers.

In the second approach, the middleware components will be virtualized into a virtual layer which will be the integration point providing management and unified access to the three main middleware services (discovery, management and communication) for IoT devices. Both approaches will be experimentally evaluated to determine their suitability and effectiveness in the different scenarios in which the integrated heterogeneous IoT platforms usually operate. At this level, reliability, real-time requirements and security (and trust) need to be guaranteed by defining suitable policies and algorithms, and incorporating them into the overlay middle components (mediator and broker) or into the virtualization layer.

Table 4 IoT middleware main features

IoT middleware feature	Source	IMPReSS	ACOSO	UBIROAD	HYDRA
Device abstraction - Heterogeneous devices need to be abstracted in virtualized, homogeneous entities in order to couple them or make them interact	Rl, R2, R3	T	T	T	T
Hardware/software interface abstraction - Interfaces need to be made generic and standardized through higher level mechanisms so that their use will be straightforward	Rl, R2, R3	T	T	T	T
Heterogeneous data source and type management - Data generated according to different modalities, formats and types require shared representation to be exchanged and exploited	Rl, R2, R3	T	T	T	T
Device Management - Device need to be efficiently and autonomously discovered, used and composed, trying to minimize the human intervention	Rl, R2, R3	P	T	T	T
Context-awareness - Implicit and explicit information about users, devices, and the environment need to be considerate for enhance the service provision	Rl, R2, R3	T	T	T	T
Security and privacy - Efficient and scalable mechanisms are needed to ensure global connectivity and accessibility but, at same time, security and privacy	R2, R3	T	N	T	T
Development process support - Suitable methods and tools need to be defined to effectively and systematically support the development process	Rl, R2	P	T	N	N
Reliability and Timeliness - Specific methods need to be defined for guaranteeing the reliable and on-time delivery of information	Rl, R2	P	T	P	T

3.1.4 Application Services

There are currently very different paradigms for Cloud-based services supporting the IoT. They range from Virtual Objects [51], which mirror a sensor or 'thing' in the network with its abstract representation in the cloud, to smarter yet more complex multi-agent [31] and event-based architectures [45]. These approaches allow the definition of logical paths for the data, based on internal and external information including aggregation. Additionally, there are a number of meta-services, such as service discovery, management, and live-updates, which facilitate the deployment and functioning of a heterogeneous IoT system [25].

State-of-the-art communication network architectures and solutions for the real-time information exchange are characterized by the adoption of virtualization technologies, SDN in data centres and cloud virtualization, respectively, and service architectures developed in the M2M domain clearly separate functional entities and service layers in the device and network domains. While standardization (e.g., by oneM2M.org) is still on its way toward stable and widely accepted specifications, commercially available service platforms - offered as licensed software on customer infrastructure or software as-a-service – already drive down cost of solution development, through powerful horizontal services, and enable cost-efficient scalability in the service delivery [34].

Regardless of the abstraction level, what they all have in common, is that they all run in a network designed for high bandwidth and short delays, which is foreign to the IoT network that is designed for efficiency and low consumption, and where intermittent communication failures are expected. Cloud services for IoT have a wide spectrum including data storage, information synchronization, data analysis, and M2M communication, as well as others that are more specific, such as geolocation or streaming. Often, they are not IoT specific, and can greatly benefit from a layer that manages the interaction between the two, making it easier for both parties to operate in an efficient way. For example, a very simple low-power sensor can make use of a very high level cloud-based service without degrading its battery life by means of interoperability at network, middleware, and some additional caching app services. That is a very powerful mechanism that allows us to connect existing IoT networks with existing Cloud Services without the need of modifying either.

INTER-IoT framework aims to be generic, allowing for different approaches to coexist on the platform like cloud services that make use of an IaaS layer to scale its functioning as needed. Nevertheless, the platform will provide the necessary components to support different approaches and meta-services, such as a de-coupling middleware system that will effectively separate the different networks in order to present an appropriate behaviour to the different elements on each side of the communication. Despite the work done in the integration and homogenization of IoT systems, most efforts aim to connect different standards and services within a single virtual environment and at a single level of abstraction. **INTER-IoT**, on the other hand, will provide with a cross-level integration schema, which allows for diverse elements to interact without the need of additional intermediaries, using NodeRed[2] as an example of such integration.

In particular, **INTER-IoT** aims to make interoperable and/or integrate application services furnished by heterogeneous IoT platforms. To fulfil it, we are defining methods based on service-oriented computing and virtualization. Specifically, application services will be first virtualized and then managed through a well-defined virtual service management component that also aims to provide automated service composition. Moreover, service composition needs to be reliable and secure.

3.1.5 Data and Semantics

Semantic interoperability can be conceptualized as an approach to facilitate "combining" multiple IoT platforms. The simplest case, of combining two IoT platforms, could be addressed by developing a one-to-one translator (a "gateway") to allow "semantic understanding" between them [33]. However, this approach does not scale, as for every subsequent entity joining an assembly of N platforms. Thus, N translators would have to be created. The two main approaches to avoid this problem, and deal with semantic interoperability are: (i) common communication standards; (ii) ontology and semantic data processing.

Developing a common communication standard was tried in the travel domain[3] with the OTA message specification a standard consisting on a set of (XML-demarcated) messages; or in the healthcare domain (and thus related to the INTER-Health use case) with OpenEHR,[4] which is an open domain-driven platform for developing flexible e-health systems. Here, multiple projects strive to establish interoperability between already known standards and the OpenEHR, e.g., establishing semantic interoperability of the ISO EN 13606 and the OpenEHR archetypes [32]. Similarly, the Think!EHR Platform (health data platform based on vendor-neutral open data standards designed for real-time, transactional health data storage, query, retrieve and exchange)[5]; aims at establishing interoperability of the OpenEHR and the HL7 standard (a framework for the exchange, integration, sharing, and retrieval of electronic health information).[6] Interestingly, development of the Think!EHR Platform had to deal with the data standards problem caused by existence of HL7 RIMv3, ISO13606, and OpenEHR standards.[7] While it is possible to envision an approach similar to this, applied to individual domains, it is not likely to be easily generalizable

[2]http://nodered.org.

[3]Open Travel Alliance available at http://www.opentravel.org/.

[4]OpenEHR available at http://www.openehr.org/what_is_openehr.

[5]Think!EHR platform, available at http://www.marand-think.com/.

[6]http://www.hl7.org/implement/standards/.

[7]Borut Fabian, "Interoperability with Think!EHR", available at http://www.hl7.org/documentcenter.

to support interactions between domains. Therefore, approaches based on ontologies and semantic data processing will be used in the project.[8]

INTER-IoT approach is developing a generic ontology of IoT Platforms (GOIoTP). The GOIoTP is used as the centrepiece for establishing platform interoperability (allowing for, among others, data interoperability, message translation, etc.). It should be stressed that, state-of-the-art ontologies of the IoT,[9] will constitute the starting point for construction of the GOIoTP, needed in our project. The proposed approach will require, (i) ontology matching [41], (ii) merging, noting that ontology merging is often reduced to ontology matching,[10] as well as (iii) techniques for establishing semantic distance (needed for ontology matching) [38]. Observing that this approach allows "understanding" and adaptability (handled through ontology adaptation) of heterogeneous data.

The creation of GOIoTP in **INTER-IoT**, combined with the state-of-the-art approaches to ontology matching/merging, allows the development of a comprehensive support for facilitation of semantic interoperability between IoT platforms, in the form of a IoT Platform Semantic Mediator (IPSM). The resulting approach, based on conducted research, will consist both of the methods and supporting tools and will include, among others, methods for:

- Combining two (or more) IoT platforms with explicitly defined ontologies. Here, among others, the following issues will be researched: (i) bringing multiple ontologies to a common format/language (for example, transforming XML into RDF and further transforming it into OWL-demarcated ontology using XLST), (ii) ontology matching, to allow for (iii) ontology merging into the extended GOIoTP (as the top-level ontology).
- Joining an "incoming" IoT platform (with an explicitly defined ontology) to an existing federation of IoT platforms (with an already defined common ontology). Here the process would be somewhat a simplified version of the previous method as only two ontologies will be integrated.
- Dealing with IoT platforms without an explicitly defined ontology/taxonomy/etc. Here, appropriate set of tools will be adapted to help instantiate an ontology for the multi-IoT-platform under construction. Specifically, the ontology will be built on the basis of information contained in one, or more: (i) definition of used data; (ii) structure of the database(s); (iii) queries issued on the database(s); and (iv) exchanged messages.

In particular, **INTER-IoT** defines methods for data and semantics interoperability. The key method for data semantics and interoperability is the development of GOIoTP (see above). Next a complete method, and a set of tools, to support development of platform semantic interoperability layer will be created. The method, resulting from research activities undertaken within the task, will include, among others, issues involved in (i) combining two (or more) IoT platforms with explicitly

[8]http://www.semantic-web-journal.net/sites/default/files/swj247_0.pdf.

[9]https://hal.inria.fr/hal-00642193/document.

[10]http://www.jfsowa.com/ontology/ontoshar.htm#s5.

defined ontologies (in any format). Here, the method will take into consideration: (a) bringing ontologies to a common format/language, (b) ontology matching, and (c) ontology merging within the GOIoTP. (ii) Joining an incoming IoT platform (with an explicitly defined ontology) to an existing federation of IoT platforms (with an already defined common ontology). (iii) Dealing with IoT platforms without an explicitly defined ontology/taxonomy/etc. Here, appropriate set of tools will be developed to help instantiating an ontology for the multi-IoT-platform under construction. Main achievements: (a) experimentally-tested methodology for IoT platform semantic integration for all possible cases of onto-semantic inputs from platforms to be integrated and (b) experimentally-tested tools for IoT platform semantic integration for all possible cases of onto-semantic inputs originating from platforms to be integrated.

3.1.6 Cross Layering

INTER-IoT specifically aims at creating cross-layer interoperability and integration between heterogeneous IoT platforms. Cross-layer approaches are fundamental to made interoperable/integrate the whole layer stack (device, networking, middleware, application service, data and semantics) of IoT platforms. Cross layering will be therefore based on the outcomes of the previous points (see Sects. 3.1.1–3.1.5).

Moreover, important requirements and features such as Quality of Service (QoS), Quality of Experience (QoE), Security, Privacy, Trust and Reliability, require to be addressed at each layer with different mechanisms. Such transversal approach allows retaining the benefits of a layered architecture (e.g., modularity, interoperability, etc.) but adding, at the same time, flexibility (e.g., optimization, tunable design, etc.) to those components that require it. Considering the heterogeneity and spread of IoT devices and IoT applications, it is straightforward that such design choice is more than suitable to properly support (i) dynamic QoS and QoE (the former, basically aiming at splitting traffic up into priority classes and trying to guarantee a particular performance metric, the latter at combining more subjective aspects related to user perception into evaluating a service) [9]; (ii) novel security and privacy techniques (that consider the cyber-physical nature of IoT devices as well as of the IoT application contexts) [39]; extended trust models (in which unconventional actors, like social networks, play an important role) [50] and (iv) enhanced reliability mechanisms (to deal with failure of resource-limited IoT device, lack of coverage from access networks in some region, rapid application context switches, etc.) [30].

3.2 INTER-FW

The Interoperable IoT Framework (INTER-FW) aims at providing global and open platform-level interoperability among heterogeneous IoT platforms coupled through specifically developed LIIs. INTER-FW will rely on an *architectural meta-model* for

IoT interoperable platforms and on a *metadata-model* for IoT interoperable semantics. Figure 3 shows the abstract schema of the INTER-FW. It provides a programming library (i.e., INTER-API) that will be used both by the INTER-FW tools, providing global-level management of the integrated IoT platforms, and by new, possibly cross-domain, IoT applications developed atop INTER-FW and that will be developed in WP4, in full compliance with the designed INTER-IoT meta architecture and meta data model.

Thus, INTER-FW advances the state-of-the-art by providing a general and effective method for inter-platform interoperability, addressing at a global level: real-timeliness, reliability, security, privacy, trust. In particular, INTER-FW will be designed and implemented considering the need to respect, where applicable, user data privacy (e.g., anonymization, hidden ID, use of separate databases for identification and data content with controlled access) and secure access to data (only authorized devices, ensure authentication and non-repudiation). Furthermore, access to data by non-authorized parties should be prevented (especially malevolent ones). Every other functional and non-functional requirement (e.g., reliability or user-friendliness) will be incorporated to the specification and implementation of INTER-FW, including the tools, INTER-API and an interoperability flexible engine. INTER-FW includes also a management mechanism and API used to access and coordinate the different layers of INTER-IoT. This aspect of INTER-FW is mainly needed for aspects like discovery, registration of devices and smart objects and also for security and privacy management.

3.3 INTER-METH

The engineering methodology INTER-METH aims at supporting the integration process of heterogeneous IoT platforms to obtain interoperability among them and allow implementation and deployment of IoT applications on top of them. To date, no proposals provide a systematic methodology driving the integration implementation (see Tables 1 and 2). It is widely recognized that using an engineering methodology is fundamental in any engineering application domain (e.g., software engineering, hardware/software codesign, civil engineering, etc.). The manual and non-systematic application of complex techniques, methods and frameworks would very likely lead to an increase of the degree of errors during integration [13]. The process of INTER-METH is shown in Fig. 4. It is envisioned as iterative, including the following six phases: Analysis, Design, Implementation, Deployment, Testing and Maintenance. Each phase produces work-products that are inputs for the successive phase/s. Iteration could involve single phases, set of successive phases or the whole process, thus assuring adaptability to new elements.

In particular:

- The Analysis phase defines the integration requirements, both functional and non-functional (e.g., real-timeliness, reliability, security, privacy, trust).

- The Design phase produces the design of the integration in terms of diagrams of (i) layer interoperability infrastructures and related interfaces, and (ii) INTER-FW programming and management patterns, to fulfill the elicited requirements.
- The Implementation phase focuses on the implementation of the design work-product to obtain the full-working (hardware and/or software implemented) system.
- The Deployment phase involves the operating set-up and the configuration of the integrated IoT platform.
- The Testing phase allows performing tests to validate the integrated platform according to the functional and non-functional requirements.
- The Maintenance phase manages the upgrade and evolution of the system.

A Computer Aided Software Engineering (CASE) tool, named INTER-CASE, for integration is under development to provide full support to the automated application of INTER-METH, covering all aforementioned integration phases.

4 Use Cases

The INTER-IoT approach is use case-driven, implemented and tested in three realistic large-scale pilots:

- **INTER-LogP**: it has been designed and built to specifically accommodate the communication and processing needs of moving vehicles and cargo items (being conceived as moving things according to the IoT paradigm), e.g., by seamless and secure integration of various vehicle telematics solutions as well as mobile devices serving as retrofitting equipment. It will work over smart containers (i.e., reefers and IMOs), trucks and different infrastructures, allowing exchange of information associated with the operations and movements of containers inside the terminal.

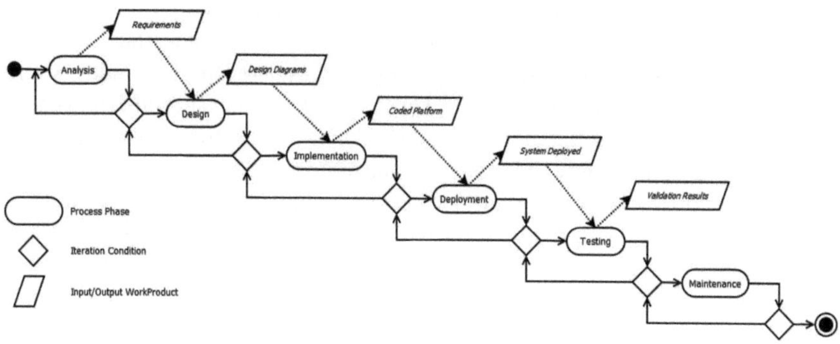

Fig. 4 Abstract schema of the INTER-METH process

- **INTER-Health**: aims at developing an innovative, open integrated m-Health IoT platform for humans monitoring in a decentralized way and in mobility. The integrated platform, derived from existing platforms (i.e., UniversAAL [24], and BodyCloud [16]), will be open to be further enhanced by integrating new subsystems by using the INTER-IoT approach.
- **INTER-Domain**: a cross-domain pilot involving IoT platforms from different application domains, including transport and logistics and e-Health, but extendable to other domains (e.g. smart cities or smart mobility).

In the following subsections, INTER-Health and INTER-LogP will be described in more detail.

4.1 INTER-LogP

4.1.1 Smart Port Transportation for Containers and Goods

In the ports of the future, port users, equipment and infrastructures will achieve a zero distance interaction offering more sustainable transport solutions. The use of IoT platforms will enable locating, monitoring, and handling different transport and cargo equipment and storage areas. The requirements for a better management of equipment and resources and the huge complexity of interactions involving large quantity of simultaneous transport movements around big logistics nodes (e.g., container terminals, ports, warehouses and dry ports) originates the need to introduce IoT platforms with multiple sensors in all logistics elements to control and monitor the several operations like energy consumption, gas emissions, or machine status. With these platforms, logistics service providers will be able to monitor and control in real time all the operations and movements of freight, equipment, vehicles and drivers on logistics nodes.

The Port of Valencia premises extend for several square kilometres. It is the largest Mediterranean port in terms of container handling. The port contains five container terminals (e.g., NOATUM and MSC), and several other facilities (e.g., train freight station, warehouses, and parking spaces). The port includes several kilometres of road within the premises.[11] The Port Authority has several deployed IoT platforms connected to different HMI and SCADA with different goals (e.g., traffic management, security, safety and environmental protection, or vessels identification). Some of these platforms provide selected data to the Port Community System (PCS) like tamper proof RFID tags and e-seals that are installed on trucks and semi-trailers. In particular, A Port Community System is an electronic platform that connects the multiple systems operated by a variety of organisations that make up a seaport, airport or inland port community. It is shared in the sense that it is set up, organised and used by firms in the same sector – in this case, a port community. There is an

[11] http://www.valenciaport.com/BoletinEstadistico/2013/December%202013.pdf.

increasing need that trucks, vehicles and drivers seamlessly interoperate with the port infrastructures and vice versa. All deployed IoT platforms do not interoperate as they are based on different standards, and remain isolated with a clear lack of interoperability at all layers.

NOATUM Container Terminal is one of the biggest container terminals in the Mediterranean located at the port of Valencia. It is the fifth largest European port in container handling, i.e. it deals with more than 50,000 movements per day, produced by more than 200 container handling units (e.g., cranes, forklifts, RTGs, internally owned tractors and trailers, etc.); more than 4,000 trucks and other vehicles visit terminal premises; with more than 10,000 containers involved in these movements. These values show the complexity of this environment and the opportunities that the information compiled by the sensors installed on the equipment, trucks and containers; and the IoT interconnected architectures can bring to the terminal (e.g., in terms of optimization in the operations, safety, security or cost and energy savings). Container terminals like the one managed by the NOATUM have a huge number of sensors, CPS (Cyber Physical Systems) and smart objects; fixed and mobile deployed and exchanging information within one or between several platforms deployed in their premises. The sensors from the internal equipment (i.e., container terminal IoT ecosystem), constitute 5% of total vehicles moving daily within terminal premises, and they generate more than 8,000 data units per second. The other 95% of the vehicles are external trucks and other vehicles, with sensors belonging to other IoT ecosystems, currently unable to interact with the terminal IoT solution. Additionally, containers (mainly used to transport controlled temperature cargoes) have their own IoT architecture, which cannot be accessed by the terminal, when the container is stored in the yard or moved across it. This lack of interoperability of outdoor ambulatory IoT things based on heterogeneous architectures represents a big barrier that **INTER-IoT** aims at removing.

This use case illustrates the need to seamlessly IoT platforms interoperation within port premises, e.g., container terminal, transportation companies, warehouses, road hauliers, port authorities, customs, border protection agencies, and outside the port. Port IoT ecosystems used to be operated by a large number of stakeholders, and typically require high security and trust, due to mobility and seamless connectivity requirements, that currently are not available with the exception of proprietary and isolated solutions. Introduction of interoperability mechanisms and tools will therefore bring about new solutions and services leading to developments of the ports of the future.

4.1.2 The INTER-LogP Pilot

INTER-LogP will be an **INTER-IoT** outcome to facilitate interoperability of heterogeneous Port Transport and Logistics-oriented IoT platforms already in place, i.e., VPF and NOATUM and other components that will be brought to the use case in order to achieve the **INTER-IoT** proposed goals, e.g., I3WSN from UPV and other IoT platforms from companies operating in the Port managed premises.

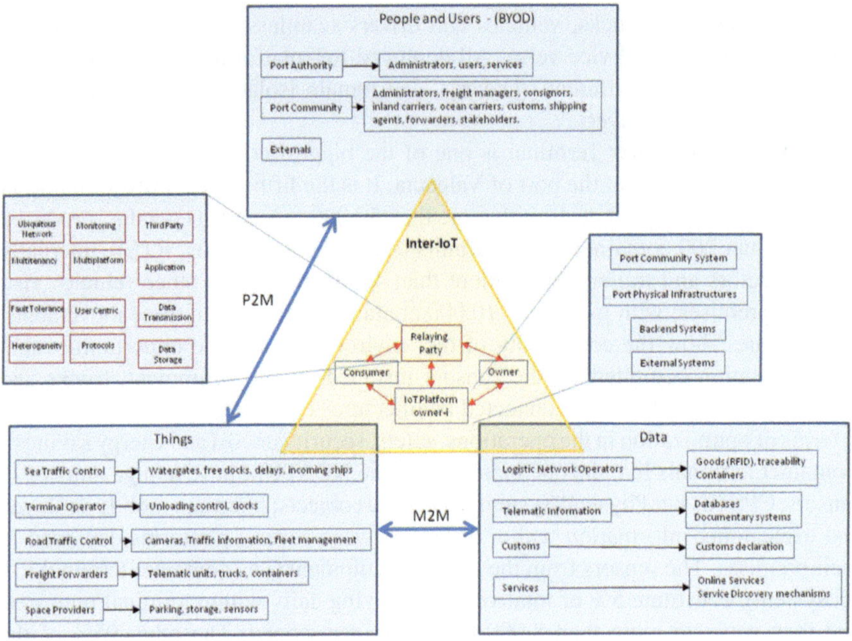

Fig. 5 **INTER-IoT** interconnection for transport and logistics (INTER-LogP)

The Port Authority of Valencia will provide its own IoT platform ecosystem to the project, including (i) the climate and weather forecast infrastructures, which monitor the environmental conditions in real-time and maintain historical data; (ii) beacon data acquisition system, which monitors and controls whenever necessary all the buoys distributed on the sea side; (iii) PCS-IoT platform, developed to cover different transportation and logistics components throughout the port premises, integrates an internal communication network and connects (more than 400) operating companies in the port (see Fig. 5).

NOATUM provides the SEAMS platform to be included in the INTER-LogP use case. SEAMS is an outcome from the Sea Terminals action (Smart, Energy Efficient and Adaptive Port Terminals) co-funded by the Trans-European Transport Network (TEN-T). It is an operational tool based on the reception of real-time energy and operative data coming from the whole machinery and vehicle fleets of NOATUM Container Terminal Valencia (NCTV). SEAMS integrates the whole set of machines (including Rubber Tyre Gantry cranes (RTG), Ship-To-Shore cranes (STS), Terminal Tractors (TT), Reach Stackers (RS) and Empty Container Handlers (ECH)) and vehicles deployed and available in the terminal premises.

INTER-IoT will help to expand the possibilities offered by not only SEAMS and the sensors installed on its own container terminal vehicles and container handling equipment units, but also sensors available on third party equipment (i.e., reefer containers)[12] and vehicles (i.e., external trucks picking up and delivering containers). Finally, it will allow installation of sensors on legacy equipment that does not have them available. Moreover, **INTER-IoT** will allow to seamlessly connecting the container terminal IoT ecosystem with other ecosystems owned by other parties, e.g., the port authority, road hauliers, the individual trucks, vehicles, containers and vessels through intelligent objects offered by different vendors, some of them managed by the PCS.

On the other hand, UPV will provide I3WSN [22], semantic IoT methodology and platform deployed in application domains like factories, automotive and defence. This generic architecture was developed within a large Spanish National project FASyS[13] and has been extended to be used in different areas like port transportation and m-health. The framework provides interoperability at different layers and includes reliability, privacy and security by design. Additionally, devices from the partners will be added to the trialsand devices from the users (e.g., truck drivers or terminal operators) like smart phones will be added to the system following BYOD (Bring Your Own Device) philosophy, allowing the integration of COTS devices in the large scale trials.

Although the different platforms that the transport and logistics use case integrates (in particular, IoT-PCS from VPF, NOATUM TOS, I3WSN UPV and the IoT platforms from other stakeholders) share some characteristics, they have different aims (i.e., focused on the particular benefits of the administrator/operator and use different technologies). All of them gather data, using different M2M and P2M protocols; some of them are cloud-based and others will be, but the most important thing is that they lack interoperability in terms of the five identified layers. There is a potential integration using one of the platforms (i.e., IoT-PCS) as a matrix architecture; however interoperability and integration will not profit the power of the proposed approach neither the capabilities of interoperable architectures rather than interconnected architectures. The use case, mainly focused in the transportation of containers, as it is the most sensorized in port transportation (especially reefer and International Maritime Organization – IMO safe containers), may improve efficiency, security and benefits to the whole transport chain. Additionally, **INTER-IoT** will provide the possibility to interact with other IoT platforms available in the port surroundings like Valencia City FIWARE infrastructure (i.e., VLCi) that is an open platform that will provide contextual information for different services and interactions at data and services layers.

[12]http://en.wikipedia.org/wiki/Refrigerated_container, last visited 13th April 2015.

[13]http://www.fasys.es/en/index.php, last visited 13th April 2015.

4.2 INTER-HEALTH

4.2.1 Decentralized and Mobile Monitoring of Assisted Livings' Lifestyle

The Decentralized and Mobile Monitoring of Assisted Livings' Lifestyle use case [35], aims at developing an integrated IoT system for monitoring humans' lifestyle in a decentralized way and in mobility, to prevent health issues mainly resulting from food and physical activity disorders. Users that attend nutritional outpatient care centres are healthy subjects with different risk degrees (normal weight, overweight, obese) that could develop chronic diseases. Only the obese (in case of second and third level obesity) need, at times, hospital care and get into a clinical and therapeutic route. The medical environment in which the pilot will be developed and deployed is the Dept. of Prevention/Hygiene Nutrition Unit at ASLTO5.

The use case will focus in the fact that in main chronic diseases, such as cardiovascular diseases, stroke, cancer, chronic respiratory diseases and diabetes, there are common and modifiable risk factors that are the cause of the majority of deaths (and of new diseases). Between the common and modifiable risk factors there are wrong lifestyles such as improper and hyper caloric diet and, in particular, the lack of physical activity. Every year in the world [48]: 2.8 million people die for obesity or overweight; 2.6 million people die for high cholesterol levels; 7.5 million people die for hypertension; 3.2 million people die for physical inactivity. These wrong lifestyles are expressed through the intermediate risk factors of raised blood pressure, raised glucose levels, abnormal blood lipids, particularly Low Density Lipoprotein (LDL cholesterol) and obesity (body mass index $\geq 30\,\text{kg/m}^2$) [43].

According to the reference standard medical protocol for the global prevention and management of obesity [46, 47], written by the World Health Organization, in order to assess the health status (underweight, normal weight, overweight, obesity) of the subject (of a given age) during the visit at the healthcare center, objective and subjective measurements should be collected (and/or computed) by a healthcare team (doctor, biologist nutritionist, dietician, etc.). The objective measurements are: weight, height, body mass index (enabling diagnosis of overweight and obesity), blood pressure or waist circumference. The subjective measurements reported by the subject, are collected through computerized questionnaires, and concern the eating habits: quality and quantity of food consumed daily and weekly, daily consumption of main meals (breakfast, lunch, dinner and snacks) and the practice of physical activity (quality and quantity of physical activity daily and weekly). The physical activity degree is detected subjectively during the first visit and could be objectively monitored through wearable monitoring devices. On the basis of these measurements, the caloric needs are automatically calculated, and the diet of the subject is defined. From this point forward, the subject must be monitored periodically (for example, every three months) for a period of at least one year. Usually monitoring is carried out at the health-care center, where the objective and subjective measurements are cyclically repeated. Based on the results, and depending on the health status reached by the subject (improved or worsened), the possibility of redefining his diet and his physical activity is analyzed.

By exploiting an integrated IoT environment, the aforementioned monitoring process can be decentralized from the healthcare center to the monitored subjects' homes, and supported in mobility by using on-body physical activity monitors. Specifically, the system will be created by using a new IoT platform, named INTER-Health [35], obtained by integrating two already-existing heterogeneous, non-interoperable IoT platforms for e-Health according to the approach proposed in the **INTER-IoT** project, based on the INTER-FW and its associated methodology INTER-METH: (i) UniversAAL, developed by UPV [24], and (ii) BodyCloud [16], developed by UNICAL.

4.2.2 The INBTER-Health Pilot

There is a need of integrating different IoT platforms as proposed in the INTER-Health use case. The effective and efficient integration of heterogeneous e-Health IoT Platforms will provide an appropriate answer to the challenges described in **INTER-IoT** proposal. The two platforms considered are UniversAAL and BodyCloud, and the result of the integration will allow developing a novel IoT m-Health system for Lifestyle Monitoring.

This flexibility allows deploying universAAL-based solutions in multiple configurations, such as local-only nodes, mobile nodes connected to server instances, or non-universAAL nodes connecting to a multi-tenant server. Communication between applications and/or sensors happens through three different buses. Messages and members are always described semantically using the domain ontologies at hand: (i) Context bus - An event-based bus for sharing contextual information from context publishers to context subscribers; (ii) Service bus - A request-based bus for on-demand execution and information retrieval from service callers to service providers and (iii) User Interface bus - A centrally-managed bus that allows applications to define abstract interfaces to be rendered by different User Interface (UI) modalities. In each bus, semantic reasoning is used to match the transferred messages to the appropriate destination. This way, applications and sensors only need to describe what they provide and what they require from others. There is no need to specify recipients, connections nor addresses explicitly [24].

BodyCloud [16] is a SaaS architecture that supports the storage and management of body sensor data streams and the processing (online and offline analysis) of the stored data using software services hosted in the Cloud. In particular, BodyCloud endeavours to support several cross-disciplinary applications and specialized processing tasks. It enables large-scale data sharing and collaborations among users and applications in the Cloud, and delivers Cloud services via sensor-rich mobile devices. BodyCloud also offers decision support services to take further actions based on the analyzed BSN data.

The BodyCloud approach is centered around four main decentralized components (or sides), namely Body, Cloud, Viewer, Analyst: (i) Body-side is the component, currently based on the SPINE Android, that monitors an assisted living through wearable sensors and stores the collected data in the Cloud by means of a mobile device; (ii) Cloud-side is the component, based on SaaS paradigm, being the first general-purpose software engineering approach for Cloud-assisted community BSNs; (iii) Viewer-side is the Web browser-enabled component able to visualize data analysis through advanced graphical reporting; and (iv) e Analyst–side is the component that supports the development of BodyCloud applications.

The two platforms, UniversAAL and BodyCloud, share some high-level characteristics while differ in objectives and technology. Specifically, they are both e-Health platforms, based on Bluetooth technology to interact with measurement devices, and based on Cloud infrastructures to enable data storing, off-line analysis, and data visualization. However, they have different specific objectives and are not interoperable from a technological point of view (at each layer and at the global level). Their specific objectives are complementary: UniversAAL is focused mainly on non-mobile remote monitoring based on non-wearable measurement devices, whereas BodyCloud provides monitoring of mobile subjects through wearable devices organized as body sensor networks. Thus, their integration will produce a full-fledged m-Health integrated platform (Fig. 6) on top of which multitudes of m-Health services could be developed and furnished. The use case of Sect. 4.2.1 will be fully deployable atop the integration of UniversAAL and BodyCloud: (i) the automated monitoring at the health-care center and the decentralization of the monitoring at the patients' homes will be supported by UniversAAL remote services; (ii) the monitoring of mobile assisted livings would be enabled by the BodyCloud mobile services; (iii) new cross-platform services will be developed for enabling complete analysis of the measurement streams coming from assisted livings.

5 Conclusions and Future Work

In this chapter we have presented the INTER-IoT systemic approach, which is being created within the INTER-IoT project together with necessary software tools and end-user applications. It will provide ways of overcoming interoperability problems between heterogeneous IoT systems across the communication/software stack, including: devices, networks, middleware, application services, data/semantics. Henceforth, reuse and integration of existing and future (even standard) IoT systems will be facilitated and made possible to obtain interoperable ecosystems of IoT platforms.

As the ecosystem of interoperable devices and services expands, so will increase the value of building new devices for and applications working within this ecosystem. This emerging ecosystem is not owned by any business or entity, but rather it exists to enable many entities to pool their resources together to create larger opportunities for all. Open interoperability delivers on the promise of open source software, enabling vendors and developers to interact and interoperate, without interfering with anyone's ability to compete by delivering a superior product and experience. In the absence

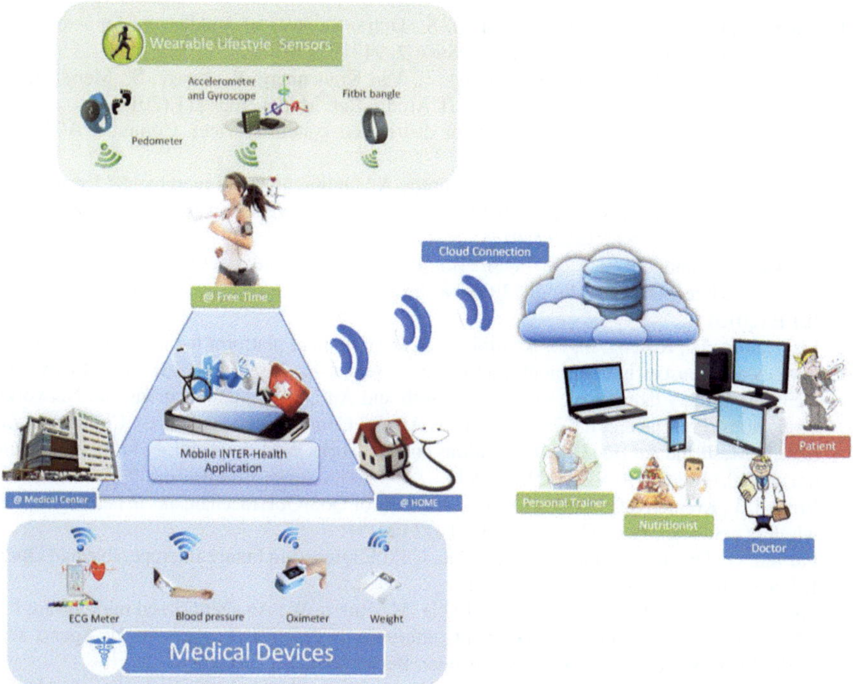

Fig. 6 **INTER-IoT** interconnection for m-Health (INTER-Health)

of global IoT standards, the INTER-IoT project and results will support and make it easy for any company to design IoT devices, smart object, or services and get them to market quickly, to a wider client-base, and to create new IoT interoperable ecosystems. In the long term, ability for multiple applications to connect to and interact with heterogeneous sensors, actuators, and controllers, thus making them interoperable, will become a huge enabler for new products and services.

Acknowledgements This work has been carried out under the framework of INTER-IoT, Research and Innovation action - Horizon 2020 European Project, Grant Agreement #687283, financed by the European Union.

References

1. Alaya, M.B., Medjiah, S., Monteil, T., Drira, K.: Toward semantic interoperability in oneM2M architecture. IEEE Commun. Mag. **53**, 35–41 (2015)
2. Aloi, G., Caliciuri, G., Fortino, G., Gravina, R., Pace, P., Russo, W., Savaglio, C.: A mobile multi-technology gateway to enable IoT interoperability. In: 2016 IEEE First International Conference on Internet-of-Things Design and Implementation (IoTDI), pp. 259–264. IEEE (2016)

3. Bandyopadhyay, S., Sengupta, M., Maiti, S., Dutta, S.: Role of middleware for internet of things: a study. Int. J. Comput. Sci. Eng. Surv. **2**, 94–105 (2011)
4. Bassi, A., Bauer, M., Fiedler, M., Kramp, T., Van Kranenburg, R., Lange, S., Meissner, S.: Enabling things to talk. Des. IoT Solut. IoT Archit. Ref. Model, 163–211 (2013)
5. Bernstein, P.A.: Middleware: a model for distributed system services. Commun. ACM **39**, 86–98 (1996)
6. Bosman, H.H., Iacca, G., Wörtche, H.J., Liotta, A.: Online fusion of incremental learning for wireless sensor networks. In: 2014 IEEE International Conference on Data Mining Workshop (ICDMW), pp. 525–532. IEEE (2014)
7. Doukas, C., Antonelli, F.: COMPOSE: Building smart & context-aware mobile applications utilizing IoT technologies. In: 2013 Global Information Infrastructure Symposium, pp. 1–6. IEEE (2013)
8. Eisenhauer, M., Rosengren, P., Antolin, P.: A Development platform for integrating wireless devices and sensors into ambient intelligence systems. In: 2009 6th IEEE Annual Communications Society Conference on Sensor, Mesh and Ad Hoc Communications and Networks Workshops. Presented at the 2009 6th IEEE Annual Communications Society Conference on Sensor, Mesh and Ad Hoc Communications and Networks Workshops, pp. 1–3 (2009). https://doi.org/10.1109/SAHCNW.2009.5172913
9. Ernst, J.B., Kremer, S.C., Rodrigues, J.J.: A survey of QoS/QoE mechanisms in heterogeneous wireless networks. Phys. Commun. **13**, 61–72 (2014)
10. Fortino, G., Di Fatta, G., Ochoa, S.F., Palau, C.E.: Engineering Future Interoperable and Open IoT Systems. Elsevier, Amsterdam (2017)
11. Fortino, G., Guerrieri, A., Lacopo, M., Lucia, M., Russo, W.: An agent-based middleware for cooperating smart objects. International Conference on Practical Applications of Agents and Multi-Agent Systems, pp. 387–398. Springer, Berlin (2013)
12. Fortino, G., Guerrieri, A., Russo, W.: Agent-oriented smart objects development, In: 2012 IEEE 16th International Conference on Computer Supported Cooperative Work in Design (CSCWD), pp. 907–912. IEEE (2012)
13. Fortino, G., Guerrieri, A., Russo, W., Savaglio, C.: Towards a development methodology for smart object-oriented IoT systems: a metamodel approach. In: 2015 IEEE International Conference on Systems, Man, and Cybernetics (SMC), pp. 1297–1302. IEEE (2015)
14. Fortino, G., Guerrieri, A., Russo, W., Savaglio, C.: Middlewares for smart objects and smart environments: overview and comparison. In: Internet of Things Based on Smart Objects, pp. 1–27. Springer International Publishing (2014)
15. Fortino, G., Guerrieri, A., Russo, W., Savaglio, C.: Integration of agent-based and cloud computing for the smart objects-oriented IoT. In: Proceedings of the 2014 IEEE 18th International Conference on Computer Supported Cooperative Work in Design (CSCWD), pp. 493–498. IEEE (2014)
16. Fortino, G., Parisi, D., Pirrone, V., Di Fatta, G.: BodyCloud: A SaaS approach for community body sensor networks. Future Gener. Comput. Syst. **35**, 62–79 (2014c)
17. Fortino, G., Russo, W.: Towards a cloud-assisted and agent-oriented architecture for the internet of things. In: Baldoni, M., Baroglio, C., Bergenti, F., Garro, A. (eds.) CEUR Workshop Proceedings WOA@AI*IA, pp. 60–65. http://CEUR-WS.org (2013)
18. Fortino, G., Russo, W., Savaglio, C.: Simulation of agent-oriented internet of things systems. In: Proceedings of 17th Workshop From Objects to Agents. pp. 8–13 (2016)
19. Fortino, G., Russo, W., Savaglio, C.: Agent-oriented modeling and simulation of IoT networks. In: 2016 Federated Conference on Computer Science and Information Systems (FedCSIS), pp. 1449–1452. IEEE (2016)
20. Ganzha, M., Paprzycki, M., Pawlowski, W., Szmeja, P., Wasielewska, K.: Semantic interoperability in the internet of things: an overview from the INTER-IoT perspective. J. Netw. Comput. Appl. **81**, 111–124 (2016)
21. Ganzha, M., Paprzycki, M., Pawlowski, W., Szmeja, P., Wasielewska, K., Fortino, G.: Tools for ontology matching—practical considerations from INTER-IoT perspective. In: International Conference on Internet and Distributed Computing Systems, pp. 296–307. Springer (2016)

22. Giménez, P., Molina, B., Calvo-Gallego, J., Esteve, M., Palau, C.E.: I3WSN: industrial intelligent wireless sensor networks for indoor environments. Comput. Ind. **65**, 187–199 (2014)
23. Gubbi, J., Buyya, R., Marusic, S., Palaniswami, M.: Internet of things (IoT): a vision, architectural elements, and future directions. Future Gener. Comput. Syst. **29**, 1645–1660 (2013)
24. Hanke, S., Mayer, C., Hoeftberger, O., Boos, H., Wichert, R., Tazari, M.-R., Wolf, P., Furfari, F.: universAAL–an open and consolidated AAL platform. In: Ambient Assisted Living, pp. 127–140. Springer (2011)
25. Iftikhar, S., Khan, W.A., Ahmad, F., Fatima, K.: Semantic Interoperability in E-Health for Improved Healthcare. Semant, ACTION-APPLICATIONS Scenar, vol. 107 (2012)
26. Kamienski, C., Jentsch, M., Eisenhauer, M., Kiljander, J., Ferrera, E., Rosengren, P., Thestrup, J., Souto, E., Andrade, W.S., Sadok, D.: Application development for the Internet of Things: A context-aware mixed criticality systems development platform. Comput. Commun. (2016)
27. Kubicek, H., Cimander, R., Scholl, H.J.: Layers of Interoperability. In: Organizational Interoperability in E-Government, pp. 85–96. Springer, Berlin, Heidelberg (2011). https://doi.org/10.1007/978-3-642-22502-4_7
28. Li, S., Xu, L.D., Zhao, S.: The internet of things: a survey. Inf. Syst. Front. **17**, 243–259 (2015). https://doi.org/10.1007/s10796-014-9492-7
29. Liotta, A.: The cognitive NET is coming. IEEE Spectr. **50**, 26–31 (2013)
30. Madsen, H., Burtschy, B., Albeanu, G., Popentiu-Vladicescu, F.: Reliability in the utility computing era: towards reliable fog computing. In: 2013 20th International Conference on Systems, Signals and Image Processing (IWSSIP), pp. 43–46. IEEE (2013)
31. Manate, B., Munteanu, V.I., Fortis, T.-F.: Towards a scalable multi-agent architecture for managing iot data. In: 2013 Eighth International Conference on P2P, Parallel, Grid, Cloud and Internet Computing (3PGCIC), pp. 270–275. IEEE (2013)
32. Martínez-Costa, C., Menárguez-Tortosa, M., Fernández-Breis, J.T.: An approach for the semantic interoperability of ISO EN 13606 and OpenEHR archetypes. J. Biomed. Inform. **43**, 736–746 (2010)
33. Mesjasz, M., Cimadoro, D., Galzarano, S., Ganzha, M., Fortino, G., Paprzycki, M.: Integrating Jade and MAPS for the development of Agent-based WSN applications. In: Intelligent Distributed Computing VI, pp. 211–220. Springer (2013)
34. Meyer, S., Ruppen, A., Magerkurth, C.: Internet of things-aware process modeling: integrating IoT devices as business process resources. In: International Conference on Advanced Information Systems Engineering, pp. 84–98. Springer (2013)
35. Pace, P., Aloi, G., Gravina, R., Fortino, G., Larini, G., Gulino, M.: Towards interoperability of IoT-based health care platforms: the INTER-health use case. In: Proceedings of the 11th EAI International Conference on Body Area Networks (BodyNets 2016). Presented at the The 11th EAI International Conference on Body Area Networks (BodyNets 2016). Turin, Italy (2016)
36. Pradilla, J., Palau, C., Esteve, M.: SOSLITE: Lightweight sensor observation service (SOS) for the internet of things (IOT). In: 2015 ITU Kaleidoscope: Trust in the Information Society (K-2015). Presented at the 2015 ITU Kaleidoscope: Trust in the Information Society (K-2015), pp. 1–7. (2015). https://doi.org/10.1109/Kaleidoscope.2015.7383625
37. Razzaque, M.A., Milojevic-Jevric, M., Palade, A., Clarke, S.: Middleware for internet of things: a survey. IEEE Internet Things J. **3**, 70–95 (2016). https://doi.org/10.1109/JIOT.2015.2498900
38. Rhee, S.K., Lee, J., Park, M.-W., Szymczak, M., Ganzha, M., Paprzycki, M., others: Measuring semantic closeness of ontologically demarcated resources. Fundam. Informaticae **96**, 395–418 (2009)
39. Roman, R., Zhou, J., Lopez, J.: On the features and challenges of security and privacy in distributed internet of things. Comput. Netw. **57**, 2266–2279 (2013)
40. Savaglio, C., Fortino, G., Zhou, M.: Towards interoperable, cognitive and autonomic IoT systems: an agent-based approach. In: 2016 IEEE 3rd World Forum on Internet of Things (WF-IoT), pp. 58–63, IEEE (2016)
41. Shvaiko, P., Euzenat, J.: Ontology matching: state of the art and future challenges. IEEE Trans. Knowl. Data Eng. **25**, 158–176 (2013)

42. Terziyan, V., Kaykova, O., Zhovtobryukh, D.: UbiRoad: Semantic middleware for context-aware smart road environments. In: 2010 Fifth International Conference on Internet and Web Applications and Services. Presented at the 2010 Fifth International Conference on Internet and Web Applications and Services, pp. 295–302 (2010). https://doi.org/10.1109/ICIW.2010.50
43. Tunstall-Pedoe, H.: Preventing Chronic Diseases. A Vital Investment: WHO Global Report. Geneva: World Health Organization, 2005. pp 200. CHF 30.00. ISBN 92 4 1563001. http://www.who.int/chp/chronic_disease_report/en.IEA (2006)
44. Vermesan, O., Friess, P., Guillemin, P., Gusmeroli, S., Sundmaeker, H., Bassi, A., Jubert, I.S., Mazura, M., Harrison, M., Eisenhauer, M., others: Internet of things strategic research roadmap. Internet Things-Glob. Technol. Soc. Trends **1**, 9–52 (2011)
45. Wan, J., Li, D., Zou, C., Zhou, K.: M2m communications for smart city: an event-based architecture. In: 2012 IEEE 12th International Conference on Computer and Information Technology (CIT), pp. 895–900. IEEE (2012)
46. World Health Organization. Obesity: preventing and managing the global epidemic. World Health Organization (2000)
47. World Health Organization. Physical status: The use of and interpretation of anthropometry, Report of a WHO Expert Committee (1995)
48. World Health Organization, others, 2013. Global status report on noncommunicable diseases 2010. Geneva: World Health Organization. Google Sch (2011)
49. Wu, M., Lu, T.-J., Ling, F.-Y., Sun, J., Du, H.-Y.: Research on the architecture of Internet of things. In: 2010 3rd International Conference on Advanced Computer Theory and Engineering (ICACTE), pp. V5–484. IEEE (2010)
50. Yan, Z., Zhang, P., Vasilakos, A.V.: A survey on trust management for internet of things. J. Netw. Comput. Appl. **42**, 120–134 (2014)
51. Yuriyama, M., Kushida, T.: Sensor-cloud infrastructure - physical sensor management with virtualized sensors on cloud computing. In: 2010 13th International Conference on Network-Based Information Systems. Presented at the 2010 13th International Conference on Network-Based Information Systems, pp. 1–8 (2010). https://doi.org/10.1109/NBiS.2010.32

Erratum to: Integration, Interconnection, and Interoperability of IoT Systems

Raffaele Gravina, Carlos E. Palau, Marco Manso, Antonio Liotta and Giancarlo Fortino

Erratum to: R. Gravina et al. (eds.), *Integration, Interconnection, and Interoperability of IoT Systems,* Internet of Things, https://doi.org/10.1007/978-3-319-61300-0

In the original version of the book, incorrect author name "Maria Ghanza" has to be corrected as "Maria Ganzha" in Preface and in chapter "Towards Multi-layer Interoperability of Heterogeneous IoT Platforms: The INTER-IoT Approach". The erratum book has been updated with the change.

The updated online version of the chapter can be found at
https://doi.org/10.1007/978-3-319-61300-0_10
https://doi.org/10.1007/978-3-319-61300-0

R. Gravina et al. (eds.), *Integration, Interconnection, and Interoperability of IoT Systems*, Internet of Things, https://doi.org/10.1007/978-3-319-61300-0_11

MIX
Papier aus verantwortungsvollen Quellen
Paper from responsible sources
FSC® C105338

If you have any concerns about our products,
you can contact us on
ProductSafety@springernature.com

In case Publisher is established outside the EU,
the EU authorized representative is:
Springer Nature Customer Service Center GmbH
Europaplatz 3, 69115 Heidelberg, Germany

Printed by Libri Plureos GmbH
in Hamburg, Germany